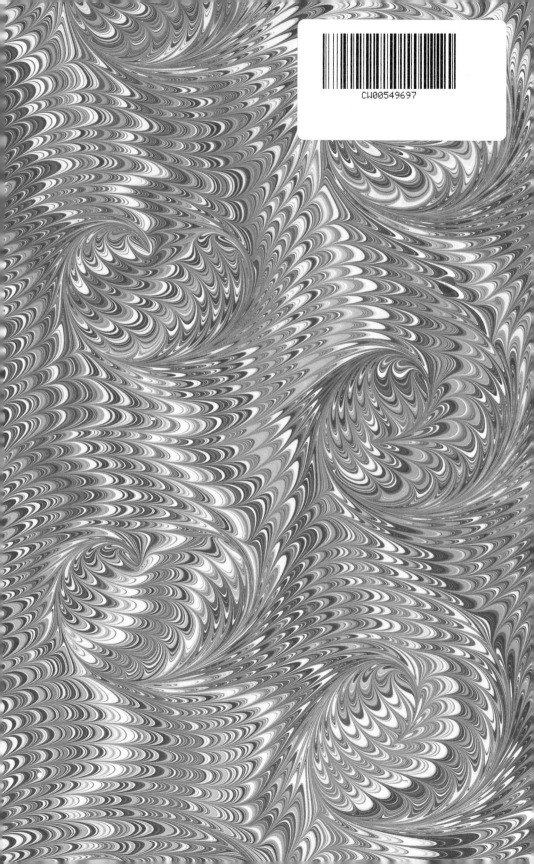

The U-2 Spyplane

The U-2 Spyplane

Toward the Unknown

A New History of the Early Years

To

Alan Johnson

Chris Pocock

Schiffer Military History
Atglen, PA

Cover: Francis Gary Powers models the uncomfortable partial pressure suit and helmet that all early U-2 pilots were obliged to wear. Known as Frank Powers to his colleagues, this quiet country boy from Virginia became a household name in 1960, after he was shot down over Sverdlovsk, captured, and put on trial by the USSR. (Lockheed U2-91-005-E or C71-2G)

Back Cover: A pair of U-2A models from the Special Projects Branch, Air Force Flight Test Center, fly over their homebase at Edwards AFB. The Branch was formed to conduct trials of an infrared sensor which could detect Soviet bombers and missiles if they attacked the US. Later, the Branch flew U-2s on a variety of research and development programs. (Lockheed PR-1187)

A U-2A in USAF markings displays the 80-foot long, high-aspect ratio wing to good effect. This aircraft was one of three which SAC's 4080th SRW deployed to Eielson AB, Alaska in early 1958 for "hot sampling" of Soviet nuclear test debris in Operation Toy Soldier - hence the tail marking. (Lockheed U2-91-028-3)

Book Design by Ian Robertson.
Copyright © 2000 by Chris Pocock.
Library of Congress Catalog Number: 99-69760

Printed in China.
ISBN: 0-7643-1113-1

We are interested in hearing from authors with book ideas on related topics.

Published by Schiffer Publishing Ltd.
4880 Lower Valley Road
Atglen, PA 19310
Phone: (610) 593-1777
FAX: (610) 593-2002
E-mail: Schifferbk@aol.com.
Visit our web site at: www.schifferbooks.com
Please write for a free catalog.
This book may be purchased from the publisher.
Please include $3.95 postage.
Try your bookstore first.

In Europe, Schiffer books are distributed by:
Bushwood Books
6 Marksbury Ave.
Kew Gardens
Surrey TW9 4JF
England
Phone: 44 (0)208 392-8585
FAX: 44 (0)208 392-9876
E-mail: Bushwd@aol.com.

Try your bookstore first.

Contents

Foreword

by Colonel Stan Beerli, USAF (retired)

In the 1950s, the USSR was a closed society which presented many unknowns to the West. Stalin had developed the atomic bomb; long-range bombers and ICBMs were being developed to deliver them. But how extensive was Soviet military power? How was it being developed, and deployed?

President Eisenhower's administration was under pressure from those who wanted to spend heavily on defense to counter the perceived Soviet threat. Some military strategists even toyed with the idea of a pre-emptive nuclear strike against the Soviet Union.

A partial intelligence picture of developments behind the Iron Curtain could be gleaned from agents, defectors, travelers, technical publications, and listening posts. But aerial reconnaissance could provide hard, undisputed evidence. The President tried to forge an "Open Skies" agreement with the Soviets, to no avail.

Some of the best minds in the U.S. addressed the problem. Richard Bissell of the CIA was considered by many to be the smartest man in Washington. Kelly Johnson of Lockheed was the dean of aeronautical engineers. The result was the U-2. Its development, operation, and exploitation may well have prevented the nuclear holocaust of World War Three. The U-2 truly represented a revolution in intelligence, and paved the way for advanced reconnaissance systems such as the SR-71 and satellites.

I was on loan from the U.S. Air Force to the CIA from 1956, to serve in the U-2 project. I checked out in the airplane and commanded operational detachments in Japan and Turkey. I then served as air operations chief in Project HQ, and subsequently as chief of the Development Projects Division. It was the opportunity of a lifetime, to be part of something which was of such vital national importance.

Chris Pocock has made a painstaking effort to research the early years of the U-2. His book is a "must read" for those who really want to understand the significance of the project in those tense, Cold War years.

Introduction

More than 20 years ago, the Lockheed Skunk Works made a short film about my favorite aircraft which ended with the following observation: "when the full story of the U-2 is cleared, it will probably confirm what military scholars and chroniclers of world events have surmised but can't yet document—that the U-2's design, production, and deployment are among the most spectacular achievements in aviation history."

I certainly agree, and have tried to document that story here in greater detail than was previously possible. However, the U-2 was part of a new reconnaissance system, in which new organizations, new sensors, and new methods of exploitation were spawned. Therefore, this book deals with more than aviation history. It explores the fascinating relationship between what was desirable from a intelligence perspective; what was technically possible; and what was politically feasible.

The U-2 was designed to uncover military and industrial developments which the Soviet Union was concealing from the outside world. To do so, the aircraft pioneered flight into a high-altitude regime that was still largely unexplored at the time. "Toward the Unknown," therefore, is an appropriate title for this book, and it derives from the legend on a cloth patch worn by U-2 pilots for many years.

My original intention was to cover the entire history of the U-2. However, the amount of new information and the constraints of time and resources have restricted my ambition. This story, therefore, ends in 1960 with the shooting-down of Gary Powers—a landmark event in the Cold War.

But, as most readers will know, the U-2 flew on. The original concept and design was so good, that nothing could fully replace it. Lockheed's own A-12 and SR-71 Blackbird spyplanes have come and gone; satellite reconnaissance has become very sophisticated; JSTARS and unmanned aerial vehicles are flying. Yet the Pentagon still keeps a fleet of some 30 much-improved U-2S models in service.

Does this study of the early years contain any lessons for those who plan, conduct, and evaluate U-2 operations today?

I think so. Each of those early U-2 flights over the Communist Bloc had closely-defined intelligence objectives; the deployments were covert in order to keep the other side guessing; the personnel involved were highly-motivated and well-rewarded; operational support from the contractors was considered essential; and the sensor "take" was analyzed by a team that was independent of the military. The U-2 is still America's most important reconnaissance aircraft, but it has not always enjoyed careful management and appropriate resources.

Chris Pocock
Uxbridge, U.K.
October 1999

1

From Boston to Burbank

On 30 November 1953, in the development planning office of the Lockheed Corporation in Burbank, California, Jack Carter wrote a memo. The recently-retired USAF Colonel titled it strategic reconnaissance, classified it "eyes only," and addressed it to his boss, Gene Root. The history of the U-2 began with that memo, in which Carter outlined the urgent need for a new type of manned reconnaissance aircraft, and some suggestions on how such a plane should be built.

The memo was unambiguous. The Soviet Union was the target, and the aircraft would have to overfly that country at extreme altitude with a payload of up to 500 pounds. Cameras would be one such payload, but Carter also mentioned "new types of sensing equipment," by which he meant radar, infrared, or electronic intelligence. The aircraft should be flown by a single pilot at altitudes between 65,000 and 70,000 feet, at a speed of mach 0.8, and have a radius of action of 1,200 nautical miles. It should be capable of avoiding virtually all Soviet defenses until about 1960.

To build such an aircraft, Carter reckoned, most of the standard rules about structural design would have to be ignored. A maneuver load factor as low as 2.25g might be required—the military norm was more like 7g. Maximum indicated airspeed should be 225-250 knots. To save weight, the landing gear should be eliminated. A turbojet which sacrificed engine life for maximum performance and weight reduction would be needed.

Why should Lockheed be bothered with such a project, since perhaps only 20 of these aircraft would be needed? Given the small numbers, Carter argued that the manufacturer might be required to contract for the whole project—development, manufacture, maintenance, operation, and personnel. If the project lasted six to eight years, the corporation might gross a worthwhile sum of between 150 and 300 million dollars from the deal, he predicted. Lockheed's reputation with the Defense Department would be enhanced, and the experience of pushing the state-of-the-art would be technically beneficial. But Carter's number one reason why Lockheed should get involved was plain old-fashioned patriotism. "The corporation would be directly contributing to the solution of one of the most vital and difficult problems facing the security agencies of the country today," he noted.[1]

Schriever, Leghorn influence

In addressing his memo to Gene Root, Carter was effectively preaching to the converted. For *both* of them had recently worked for the Pentagon office where U.S. strategic reconnaissance objectives were being redefined. This office, named USAF Development and Advanced Planning (AFDAP), set development planning objectives for various missions. It was headed by Colonel Bernard Schriever, a rising technocrat in the Air Force. Carter had been Schriever's chief-of-staff, and Root was a consultant to AFDAP from the Rand Corporation.[2]

Early in 1952, Bennie Schriever had been impressed by the ideas of Colonel Richard Leghorn, an expert and visionary in the reconnaissance field who had been recalled to military service from Kodak at the start of the Korean War. Leghorn had been drawing attention to the unique political and practical problems involved with peacetime overflights of "denied territory" since 1946. As the Iron Curtain descended on the postwar world, the requirement outlined by Leghorn and his solution—higher and faster aircraft which would be invulnerable to detection and interception—made ever more sense. Schriever invited Leghorn to join his staff at AFDAP. The young reconnaissance expert moved from Wright Field, where he had commissioned a proposal from the British manufacturer of the Canberra jet bomber, for a radical adaptation of that aircraft to the reconnaissance role.[3] At AFDAP, other staffers such as Gene Kiefer and Major Bud Wienberg agreed with Leghorn, that an aircraft designed for overflight could not be built to the usual military specifications, nor operated in the usual military fashion.

But Leghorn's proposal to adapt the Canberra had no takers in the USAF hierarchy, apart from Schriever. This was despite the advice of a group of leading scientists in fields related to reconnaissance, which had confirmed the higher-and-faster requirement in their Beacon Hill Report to the USAF in mid-1952. Leghorn was AFDAP's liaison to this study, which was named after the suburb of Boston which was close to Harvard University, MIT, and the Boston University Optical Research Laboratory, where some of the scientists worked.

Meanwhile, wartime-vintage bombers and patrol planes converted for reconnaissance missions were lumbering along—and sometimes across—Soviet borders. They faced an increasing threat of interception from the new Soviet jet fighters, such as the MiG-15, directed by air defense radars. The RB-45C reconnaissance version of America's first jet bomber, the Tornado, wasn't much of an improvement; its top speed was higher, but its max altitude was still only 40,000 feet. As the risk of detection and interception by the Soviets grew, U.S. military commanders found it increasingly difficult to gain the required political authority for deliberate overflights. A few top-priority missions over European Russia were effectively sub-contracted to Britain's Royal Air Force, since Prime Minister Churchill was more willing than President Truman to approve them.[4]

Bald Eagle Studies

Prompted perhaps by these developments and by the Beacon Hill study, the USAF formally requested design studies for a specialized reconnaissance aircraft on 1 July 1953. At the Wright Air Development Center (WADC) in Ohio, new develop-ments chief Bill Lamar and engine specialist Major John Seaberg had laid the ground-work for this request in the preceding six months. They did not, apparently, refer back to the "unconventional aircraft" work that Leghorn had done there in 1951. But it was their idea to by-pass the larger aerospace companies (including Lockheed). They thought that this relatively small requirement would get more attention from smaller companies, so they chose Bell and Fairchild. Their superiors at the Air Research and Development Command (ARDC) in Baltimore agreed. In addition, Martin was asked to study a modification of its B-57 light jet bomber, with a larger wing and new engines.[5] The three companies were asked to submit their studies by the end of the year. The codename for the project was Bald Eagle.

But when Lockheed's Jack Carter made a return visit in late 1953 to his old colleagues in the Pentagon, he found them unimpressed with the ARDC approach. Leghorn had returned to civvy street with Kodak by now, but Kiefer was still at AFDAP, pushing the idea of an unconventional airframe. Lamar and Seaberg of WADC had called for an aircraft which could reach 70,000 feet and have a radius of 1,500 nautical miles. They conceded that certain standard military equipment, such as defensive armament and an ejection seat, would have to be eliminated. They hoped that a combination of low wing loading and fast-improving turbojet engine technology would meet the requirement, rather than an utterly unconventional air-frame. To provide that turbojet, Seaberg favored Pratt & Whitney, where Perry Pratt had designed the highest pressure-ratio engine yet available. It was desig-nated the J57, and it had already been adopted for the B-52 bomber and the F-100 fighter. The J57 was the first 10,000 lbs thrust jet engine, and every pound of that thrust would be needed at 70,000 feet, where only some seven percent of an engine's sea-level thrust could be reproduced.

The Skunk Works

Carter returned to Burbank and wrote his memo to Root. Both newcomers to Lockheed, they soon became aware that the ideal group to work on an unconven-tional reconnaissance plane was right on their doorstep in Building 82. So the ini-tiative was passed to Kelly Johnson and his Skunk Works.

The Skunk Works had been created in 1943 when Lockheed got the contract to build the P-80, America's first jet fighter. Johnson was a brilliant aeronautical engineer, who had managed to cream off the pick of the Burbank factory's engineering talent into an experimental department where design engineers, mechanics, and assembly workers would work closely together in a streamlined fashion, free from the constraints imposed by the wider company bureaucracy. The department was completely independent from the rest of Lockheed for purchasing and all the other support functions. Walled off from the rest of the plant, and only accessible to the select few staff, it was also a very secure method of building secret prototypes—it built the XP-80 in just 143 days. What on earth were Kelly and Co up to in there? A popular wartime comic strip featured a hairy and eccentric Indian who regularly stirred up a big brew, throwing in skunks, old shoes, and other unlikely raw material. With Johnson's secret shop situated right next to the plastics area, the Skunk Works nickname caught on. The official name for Johnson's operation was Advanced Development Projects (ADP).

In early 1954, the Skunk Works had just completed designing the XF-104, soon to be named the Starfighter. Having heard from U.S. pilots in Korea that speed and altitude were the paramount requirements if you wanted to hassle with MiGs, Johnson had sat down to design a Mach 2 plus hotrod fighter capable of reaching over 60,000 feet. To achieve this with the jet engines then available, had meant aiming for a gross weight as low as 15,000 lb. Through ruthless pruning of systems and the design of a very short, thin wing, this had been achieved.

The CL-282

Johnson realized that the reconnaissance requirement outlined by Carter could be met by modifying the XF-104 fuselage, and marrying it to a new high-aspect, low-thickness ratio wing. He set engineers Phil Coleman, Henry Combs, and Gene Frost to work.[6] By 5 March 1954, Johnson and his small team had designed the CL-282 high altitude aircraft. It promised a maximum altitude of 73,000 feet, which would be reached at the end of a cruise-climb from 65,000 feet. The radius of action would be 1,400 nautical miles, measured from the start of cruise-climb. Endurance was therefore more than seven hours. And yet this machine took off at a gross weight of only 13,768 lbs, including a 600 lbs allowance for payload.[7]

Some key ideas that Carter had relayed from AFDAP, such as no undercarriage and low maneuver load factor/IAS, were embraced and developed by Johnson and his small team. The CL-282 would take off from a ground cart, and land on a skid attached to the lower fuselage. Otherwise, the fuselage was that of the XF-104 (minus a 62-inch forebody section) and so was the tail. But the engine was changed, from the General Electric J79 turbojet which Johnson had chosen for the F-104,[8] to the same company's J73-X52. This was an interim engine derived from the first generation J47, which powered the B-47 and the F-86. Johnson noted that it would provide greater thrust at altitude than two alternatives that he also considered (the Wright TJ31B1 and the Rolls Royce Avon RA.14).[9]

The new wing was the key to the CL-282's promised performance. It bolted to the fuselage at ring-frames, with no carry-through structure, just like the XF-104. Wingspan was 70 feet, total area 500 sq feet, and aspect ratio was 10. Although a conventional two-cell construction, this wing boasted a Kelly Johnson innovation which he originally named Span Load Distribution Control. When flying at higher speeds, or in turbulent air, the wing control surfaces could be raised to the gust position (4-degrees for the flaps and 10-degrees for the ailerons). The effect was to reduce bending moments and tail loads by moving the wing center of pressure inboard. This allowed a much lighter wing structure, but one which could still cope with relatively turbulent air. This device—soon to be renamed Gust Control—would later be utilized extensively in transport aircraft.[10]

Rival Contenders

The CL-282 design study was sent to Bennie Schriever in the Pentagon in early March 1954. He was very interested, and requested a specific proposal from Johnson. A month later, Johnson was in the Pentagon, promising to take complete responsibility for the program, including the production and field support of 30 airplanes. Two senior civilian officials at that meeting gave him a good reception, but the four USAF Generals present were not so enthusiastic.

They included Lt Gen Donald Putt, the outgoing commander of ARDC. By early April, Putt's staff at Wright Field had completed an evaluation of the industry studies which had been generated by the Bald Eagle request. Following John Seaberg's suggestion, all three contractors had adopted the Pratt & Whitney J57-P37 high altitude engine, and its development was already underway.

Martin had designed a larger wing for the B-57, and replaced its two Wright J65 turbojets with the J57s. Various weight-reducing measures were proposed, including honeycomb-backed skin panels on the wings, reduced spars and wing carry-through structure, and elimination of the flaps. Although this modified B-57 couldn't reach 70,000 ft, it would nevertheless be a quick and relatively cheap interim solution. Bell also built their Model 67 proposal around two wing-mounted J57s; theirs was a breathtakingly delicate and spindly machine with a super-slim fuselage for maximum drag reduction and wings with an incredibly high aspect ratio of 12 for maximum lift. They were nearly 115 feet long, nine feet longer than the more conventional airfoil that Martin was proposing. Having decided that a single J57 could do the job, the Fairchild team opted for a much smaller wing, but their M-195 design featured a novel engine installation above and behind the cockpit. Weighing in at a little under 11,000 lbs empty, it was less than half the weight of the Bell.

But Seaberg and his colleagues at WADC could not bring themselves to approve a single-engine design. They picked Bell and Martin instead, and after brief-

ing commanders at ARDC, SAC, and HQ USAF during April, they received immediate approval to proceed with the modified B-57.[11]

Meanwhile, the AFDAP staffers in the Pentagon tried to raise wider interest in
Kelly Johnson's plane. They sent his proposal to ARDC for formal evaluation, and
briefed SAC commander General Curtis LeMay on their reconnaissance concepts
in general, and the CL-282 in particular. LeMay perceived the AFDAP concept
(which included the idea of divorcing overflight operations from usual military
lines of command) as a threat to SAC's control of the strategic reconnaissance
mission. He stomped out of the briefing, declaring "This is a bunch of shit! I can do
all of that stuff with my B-36!"[12]

In early June 1954, HQ USAF wrote to Kelly Johnson rejecting the CL-282
because it was too unusual, only single-engined, and they were already committed
to the Martin program.[13] This was apparently before ARDC had finished its formal
evaluation of the CL-282. But Seaberg's group at Wright Field reached a similar
conclusion a few weeks later.

Original Thinking
And there the matter might have ended. The U-2 would never have been built, had
it not been for a small group of original thinkers: a key civilian in the USAF's
leadership; a leading CIA intelligence analyst; and that same group of scientists in
the Boston area who had produced the Beacon Hill report. The key USAF civilian
was Trevor Gardner, who had been personally appointed as Assistant Secretary of
the Air Force for R&D by President Eisenhower. Gardner believed that the U.S.
government should be far more concerned than it seemed to be over the threat of a
surprise Soviet attack. He had been at the Pentagon meeting when Johnson presented the CL-282, and was impressed. Gardner also knew the business, since he
was head of reconnaissance camera company Hycon in Pasadena, California, before coming to Washington.

The leading CIA analyst was Philip Strong, the operations chief within the
Office of Scientific Intelligence (OSI). Working on the front line of the U.S. effort
to gain knowledge of Soviet military developments, Strong was only too aware of
how little was known. For instance, a top-secret National Intelligence Estimate
(NIE) on Soviet guided missile capabilities was full of uncertainties: "we have no
evidence to confirm or deny current production...firm evidence on the present status of the Kapustin Yar test range is lacking...current intelligence on the particular
missiles under development is almost non-existent."[14] And missiles were not the
only problem; reports of a Soviet long-range jet bomber had first landed on Strong's
desk the previous year. On 1 May 1954, the new bomber flew over Moscow during
the Mayday parade, prompting public concern in the U.S. about the growing strategic threat.

The group of scientists included Jim Baker from Harvard University Observatory, Allen Donovan from Cornell Aeronautical Laboratory, and Edwin Land, the founder of the Polaroid company. Baker was the nation's leading lens designer, and had played a key role in reconnaissance camera development going back to WWII. Donovan was a bright aeronautical engineer who had refined Dick Leghorn's idea for a specialized aircraft during the Beacon Hill study. Land was a widely respected and dynamic pioneer of many optical devices, as well as the inventor of the Polaroid instant camera. He had first been exposed to the airborne recon business as a member of the Beacon Hill study group.

Gardner and the scientists

In mid-1953, Baker, Donovan, Land, and a few others had been invited to serve on an Intelligence Systems Panel (ISP), part of the USAF's Scientific Advisory Board. Phil Strong became associated, since he was brought in to advise the ISP of the yawning gaps in U.S. intelligence. Since he was chairman of the ISP, the USAF asked Jim Baker to visit some aircraft companies and seek their views on high-altitude aircraft. Strangely, though, Baker was not asked to visit Bell, Fairchild, *or* Lockheed. Baker subsequently paid a long visit to Europe in early 1954, where he learned first-hand of operational reconnaissance problems from USAF and RAF commanders.[15]

In mid-May 1954, Trevor Gardner and two colleagues invited Strong to the Pentagon, and briefed him on the CL-282 and the Bald Eagle designs. They asked him if the CIA would throw its weight behind the project. Strong discussed the designs with the ISP. Donovan thought the B-57 would be far too heavy; he strongly believed that a single-engine design was vital. Baker suggested that Donovan should visit Lockheed to learn more about the CL-282. Donovan wasn't able to make the trip until early August 1954, when he was briefed by Root and Johnson.[16]

By that time, the impatient Gardner had managed to energize the Eisenhower administration into commissioning a top-level panel of experts to study the surprise attack issue. It became known as the Technological Capabilities Panel (TCP), and James Killian, the president of MIT, was chosen as its chairman. As an offshoot of the President's own Science Advisory Committee, the TCP members and staff worked in the Executive Office. The TCP reported directly to Eisenhower, and had great authority. Killian subdivided the task into three projects; offensive, defensive, and intelligence capabilities. He invited fellow Bostonian Din Land onto his steering committee, and asked him to head up the intelligence project within the TCP. In turn, Land selected this project's membership, including Jim Baker and physics Nobel Laureate Ed Purcell from Harvard.[17]

As soon as Land arrived in Washington on TCP duties, Phil Strong approached him with drawings of the CL-282, and told the story of how it had been rejected by the USAF. Strong was lobbying for a flight over Kapustin Yar at this time, but the USAF said it couldn't be done. CIA director Allen Dulles wasn't very supportive, either. Land intuitively agreed with Strong's frustration. He believed in taking "a clean fresh look at the old, old knowledge." Land saw that Johnson had done just that with the CL-282. He also saw the merit of Leghorn's idea from way back, that such an unconventional plane should also be operated in an unconventional way. Land's innovative mind began to consider alternatives to the standard modus operandi of the military services.[18]

Two engines - or one?

In early September, Donovan and Baker paid a visit to Bell Aircraft's Niagara Falls facility to discuss the Model 67.[19] That same month, ARDC designated this aircraft as the X-16, and issued a contract for 28 aircraft. But Donovan was unimpressed. He discounted the safety factor that was implied by twin engines. If one engine failed at high altitude during the mission, he noted, the cruise could only be maintained at a lower altitude. There, Donovan argued, the aircraft would be easy prey for Soviet fighters. Donovan may also have been concerned about aerolastic divergence, which was a significant risk on the X-16, with its huge wing. Moreover, the Bell design was heavy with unnecessary extras, such as reinforced cockpit structure for pilot protection. Donovan still preferred the CL-282, which promised the same or better performance at almost one-third the gross weight.[20]

Land was now lobbying aggressively for the Lockheed plane, and also consulting widely to ensure that it carried the best possible sensor systems. He encouraged Jim Baker to pursue his innovative ideas for reducing camera size and weight, and brought in Richard Perkin of Perkin-Elmer for further advice. Land also persuaded Kodak to continue development of a new plastic film base named Mylar. Because it was so much thinner than existing film bases, a sufficient quantity for long-duration flights could be carried without too great a weight penalty. Land's panel also investigated electronic sensors, with a view to equipping a small high-flyer with a useful signals intelligence (SIGINT) payload.

In October 1954, Baker was sent to Burbank for a detailed discussion on the camera payload weights and bulk. Kelly Johnson fought to keep down both. Every pound added meant two feet of altitude lost, he reminded Baker. The lensman had three different camera configurations in mind: a conventional trimetrogen mapping camera; multiple reconnaissance cameras with focal lengths from 12 to 48 inches in rocking mounts; and a 180-inch design with folded optics for the most detailed coverage of specific targets. The latter was particularly difficult to accommodate in the CL-282's small fuselage. But Baker was offering the aircraft designer one very useful weight reduction. He had contacted Eastman Kodak, and they promised a new emulsion on mylar, a much thinner film base. It could be wound more tightly in the film magazine, allowing sufficient length to be carried for very long photographic missions.[21]

Land knew that one particular USAF objection to Johnson's plane was the engine. He summoned Perry Pratt and asked him to describe P&W's high-altitude version of the J57. Pratt told how they were modifying the J57's alternator, oil cooler, hydraulic pump, and other key parts. But Johnson was most reluctant to give up the single-shaft GE powerplant, which was almost 1,000 lbs lighter. He knew that changing to the J57 would entail a complete redesign of the CL-282 fuselage, and abandoning the plan to quickly produce it on existing XF-104 jigs.

The CIA as sponsor

By late October 1954, Land's group had not only concluded that the radical CL-282 design must be built, but had also made the equally radical decision that its development and operation should *not* be entrusted to the USAF. Together with Trevor Gardner, they met Director of Central Intelligence (DCI) Allen Dulles and suggested that he should sponsor the aircraft. Dulles was dubious; he saw the CIA as an espionage and analysis agency which should not become involved with a major technical development effort such as this. On 5 November, Land wrote to Dulles on behalf of TCP project 3, insisting that "overflight is urgent and presently feasible." With typical chutzpah, Land urged Dulles to exercise the CIA's "right" to pioneer scientific techniques for collecting intelligence. The military should not "engage directly in extensive overflight," a task more suited to the CIA which, "as a civilian organization [could] undertake (with the Air Force's assistance) a covert program of selected flights."

Land further noted that Lockheed had offered a comprehensive support package to accompany its proposed "jet-powered glider," and promised to have four of them ready for deployment within 17 months. He pressed for an immediate go-ahead, since "the opportunity for safe overflight may last only a few years, because the Russians will develop radars and interceptors or guided missile defenses for the 70,000 foot region."[22]

Although the TCP was not due to report to the President for another three months, James Killian agreed with Din Land on the urgency of the overflight requirement. The pair went to see President Eisenhower, who asked many hard questions, but endorsed the Land group's approach.[23] Ike had learned the value of aerial photo reconnaissance in WWII, and was ready to approve any reliable means of closing the Soviet intelligence gap.

Events now moved quickly. On 18 November General Donald Putt, who had moved up from ARDC to become Deputy Chief of Staff for Development, brought John Seaberg to the Pentagon from Wright Field to brief the TCP scientists on the USAF perspective. Seaberg grudgingly conceded that the Lockheed design was "aerodynamically close" to the Bell and Fairchild designs that he had sponsored. But he insisted that the J73 "would not be good enough to do the job in Kelly's airplane."[24]

Summoned by Trevor Gardner, Kelly Johnson arrived the next day to meet General Putt in the presence of Land's group. Johnson dispelled any lingering doubts the scientists may have had. He reluctantly agreed to re-engine his CL-282 with the J57 engine, and to rethink the lack of a landing gear. At lunch the same day with Secretary of the Air Force Harold Talbott, DCI Dulles and his deputy Lt Gen Pierre Cabell, Johnson was asked how his company could commit to such a tight timescale, when others dared not. "He's already proved it three times on previous projects," Putt intervened. The group told Johnson to go ahead, and emphasized the need for extreme secrecy.[25]

Johnson flew back to California, and gained the approval of Lockheed president Robert Gross. While Johnson worked on the redesign, Dulles called a meeting of the U.S. Intelligence Board to ratify the project.[26]

At 8 am on 24 November 1954, Dulles, Cabell, Putt, and Talbott, together with the Secretaries of State and Defense, assembled before President Eisenhower to seek his formal authority to proceed. Ike gave the go-ahead, and told them to report back when the plane was ready for operations. Amazingly, that would be just 17 months later, exactly as Kelly Johnson had promised.

Notes:

[1] Lockheed memo from J.H.Carter to L.E.Root, 30 November 1953

[2] Bud Wienberg interviewed by Cargill Hall, March 1995

[3] author's interview with Richard Leghorn

[4] for a comprehensive review of such US and British missions in the early to mid-fifties, see a long article by Cargill Hall, "Cold War Overflights: Military Reconnaissance Missions Over Russia before the U-2" in Colloquy, April 1997, Vol 18, Number 1. More detail on the RAF flights, and on SAC RB-47 flights in 1955-56, can be found in Paul Lashmar's substantial book, "Spy Flights of the Cold War", Sutton Publishing, UK 1996

[5] letter from John Seaberg to NASA historian John Sloop, 25 June 1976. Sloop subsequently incorporated Seaberg's comments in his NASA History Report SP-4404, "Liquid Hydrogen as a Propulsion Fuel 1945-1959", 1978. In his letter, Seaberg claims that he and Lamar originated the concept of a specialized reconnaissance aircraft in late 1952. However, the weight of evidence suggests otherwise, including the fact that Wienberg and Leghorn both served at WADC in 1951-52.

[6] Kelly Johnson, "Log for Project X". Hereafter Johnson Log. This is Johnson's personal diary of progress on the U-2 project. A sanitized version of this diary has been released by Lockheed, but all references in this book are to the unexpurgated version which was made available to the author by another source. The first entry in the log (December 1953) suggests that Johnson may have been aware of the new requirement before Carter and Root approached him. Robert Amory (CIA Deputy Director for Intelligence at the time) states in an oral history interview conducted for the JFK Library in 1966, that Johnson was tipped off by the CIA's Philip Strong (see later this chapter). The author has assessed this evidence, and considers it unlikely.

[7] Lockheed Report LR-9732. The full report is reproduced in Jay Miller, "Lockheed's Skunk Works", Aerofax, US, 1993

[8] the XF-104 flew with an interim Wright J65 engine, however

[9] Miller "Skunk Works" book

[10] patent filing by Kelly Johnson for Span Load Distribution Control on 6 September 1955, provided to the author by Lockheed Martin Skunk Works. Ironically, Johnson may not have had an exclusive on gust control. The Fairchild M-195 design submitted to ARDC in the Bald Eagle competition (see later this chapter) featured a similar device.

[11] Seaberg letter, as above

[12] Wienberg interview, as above. On 8 May 1954, a SAC RB-47E was attacked by MiG-17s during a deliberate daylight overflight of the Kola Peninsula. Despite LeMay's bravado, not even his latest jet bomber could properly perform this type of mission.

[13] Johnson Log, 7 June 1954.

[14] NIE-6-54, "Soviet Capabilities and Probable Programs in the Guided Missile Field", released 5 October 1954, declassified in 1993.

[15] Jim Baker interviewed by author, 1995. In the UK, the RAF told Baker that they had attempted to get photographs of the Kapustin Yar missile test site from a Canberra some months earlier. In his oral history interview (see footnote 5), DDCI Robert Amory relates how this aircraft had been intercepted by Soviet fighters and nearly shot down. Despite extensive research by Lashmar, see note 4, the full story has still not emerged.

[16] Greg Pedlow and Don Welzenbach, "The CIA and the U-2 Program", CIA History Staff, 1998, p14-16, p24. This is a declassified but still partially redacted version of a history of the CIA's U-2 and A-12 programs, which was written in 1989-90 and published with a Secret classification in 1992. In this account, the authors relate that Strong carried details of the CL-282 from the Pentagon to Richard Bissell in May 1954. They cite an interview with Bissell as one of their sources for this. Although Bissell states in all of his published writings, that he was not aware of the project until just after it was given the go-ahead in late November 1954, the best evidence suggests otherwise.

[17] The other members Land chose were mathematician John Tukey from Princeton; nuclear chemist Joseph Kennedy; and Allan Latham, a former Polaroid colleague of Land who was now with the Arthur D. Little consultancy.

[18] Leghorn interview, plus Don Welzenbach, "Din Land: The Patriot from Polaroid" in Optics and Photonics News, US, 1994. This is an excellent account of Land's role in the birth of the U-2.

[19] Baker interview

[20] Pedlow and Welzenbach, p25-26. However, the assertion by these CIA authors that only the CL-282 was designed to low (eg non-milspec) load factors is incorrect. According to Jay Miller, "The X-Planes", Aerofax/Orion Books, US, 1988, which has a detailed description of the X-16, Bell's design had a maneuvering load factor of 3g, only 0.5g more than the CL-282. Pedlow and Welzenbach are also mistaken in their belief that the USAF insisted on defensive armament. See the "Design Study Requirements" for the Bald Eagle project, also in Miller's book.

[21] Baker interview

[22] "Memo for the DCI: A Unique Opportunity for Comprehensive Intelligence", 5 November 1954. This three-page memo was accompanied by a two-page summary and a covering letter.

[23] James Killian, "Sputnik, Scientists and Eisenhower", The MIT Press, US, 1977, p82.

[24] Seaberg letter

[25] Johnson Log, 19 November 1954

[26] Johnson Log, 19-23 November 1954

2

Breaking the Mold

Within hours of gaining Presidential approval, DCI Dulles turned the new project over to his Special Assistant for Planning and Co-Ordination, Richard M. Bissell. "RMB" turned out to be an excellent choice; a brilliant economics professor from Yale and MIT whom Dulles had persuaded to join the CIA the previous February. Although his background was pure East Coast establishment, Bissell had an informal and unorthodox style of management which suited similar spirits like Gardner and Johnson. Although not a scientist or engineer, he was able to quickly assimilate and evaluate technical matters. Din Land's group of civilian experts continued as important advisors. Assisted by Herb Miller, an Agency specialist on Soviet nuclear programs, Bissell quickly organized the Development Project Staff (DPS) in one of the old wartime buildings on "the little hill" near the Lincoln Memorial. The DPS reported directly to Dulles, and even within the CIA, its existence was kept on a strict need-to-know basis. The same rules applied at the Pentagon, where R&D chief Don Putt selected Colonel Ozzie Ritland as the USAF's liaison officer to DPS.[1]

On the first working day after Thanksgiving 1954, Kelly Johnson nominated Dick Boehme as his chief assistant for the new project, and together they chose 24 engineers for the detailed design work.[2] The engineers were put on a 45-hour week, and instructed not to tell anyone what they were doing—not even their closest relatives. Johnson had promised to have the first aircraft flying in eight months, but had also been pressured into a significant redesign. Compared with the CL-282, the airframe would have to be scaled up. In consequence, the gross weight would rise, and threaten the altitude performance. Johnson said he would trade his grandmother for a ten pound saving in weight! In the Skunk Works, pounds became known as "grandmothers" as his designers struggled to eliminate every unnecessary ounce.

In mid-December, Johnson again flew East. In a series of meetings in the Pentagon, a who-does-what agreement was thrashed out. Lockheed's offer to assume

21

total project responsibility was not taken up. The CIA would contract separately for the aircraft, the payloads, and some specialized systems, and the USAF would divert the required number of J57 engines. The CIA would be responsible for communications and security, the USAF for pilot training and logistics. An existing agreement, whereby military personnel were assigned for tours with the CIA's Air Maritime Division for secret covert operations, was extended to cover the new project.[3]

Johnson presented his estimate of $22.5 million to produce, flight test, and support the first six airplanes. He then flew on to the Pratt & Whitney plant at Hartford, CT, where chief engineer Bill Gwynn and his assistant Wright Parkins briefed him on the -37 version of the J57 which they were developing for the RB-57D and the X-16. Johnson bluntly told them that its fuel consumption was too high, and its thrust at 70,000 feet—which would be only seven percent of the sea-level value—was too low. In exasperation, Johnson suggested towing his aircraft partway to altitude.[4] But then Perry Pratt promised to make further improvements in a second high-altitude version of the J57 specifically for the Lockheed plane. It would have higher thrust and lower weight, and be designated the J57-P-31.

Johnson hurried back to Burbank for the initial wind tunnel tests. He pushed his engineers hard. Each specialty was handled by only one or two men, who were trusted to make the right decisions. Under the pressure of time, normally distinct phases of design, tooling, and fabrication were to be merged if necessary. When data was not yet available, inspired estimates would be made.[5]

Defining the Article

In early January, Johnson finally came up with a weight-saving design for the landing gear. The landing skid was abandoned, but not in favor of a conventional undercarriage. Instead, a bicycle-style single main gear and small tail gear would be provided on the centreline. To maintain balance on the ground, compressed rubber wheels on sprung steel legs would protrude from the wing. They would drop out upon takeoff. For landing, a skid was added to each wingtip to protect the airfoil as the aircraft came to rest. The total weight was 250 lbs, only one-third that of a conventional gear for an aircraft of this size.[6]

By the end of January, the major drawings were finished. The emerging aircraft was a clever blend of innovation and conservatism. Construction would be mostly aluminum, with the bare minimum of stringers and other forms of structural stiffening, and the skin was wafer-thin—only 0.020 of an inch around the nose, and nowhere thicker than 0.063. Skin panels were to be flush riveted, and the control surfaces would be aerodynamically balanced with virtually no gap at the hinge point. To save weight in the hydraulic system, none of the primary flight controls were boosted.

Ed Baldwin led the fuselage design. Bifurcated intakes fed the single engine, which Baldwin located at the wing mid-chord for balance. That imperative forced him to place Kelly's single main landing gear between the inlets, further forward than desirable, and posing a significant future challenge for pilots seeking to land the aircraft safely. Ahead of the main gear was a pressurized area for the reconnaissance payloads which was known initially as the equipment bay, and later as the "Q-bay." Ed Martin devised the system whereby these payloads would be attached to pallets, the pallets mounted on interchangeable hatches, and the hatches fitted flush to the lower fuselage. Another large detachable hatch fitted the top of the payload bay; when both hatches were removed a huge gap in the fuselage was apparent, since the cockpit and nose section was supported only by the mid-fuselage spar.[7]

The cockpit was the only part of the fuselage which retained the XF-104 structure. It was therefore very small, and dominated by a large control yoke more typical of a bomber or transport aircraft. The control cables were directly linked to this large yoke, on which was mounted a half-wheel to control the ailerons. The otherwise conventional instrument panel was dominated by the hooded, six-inch display from a driftsight, which Perkin Elmer would provide. Basically a downward-looking periscope, this driftsight allowed pilots to view the scene beneath their aircraft. It had a scanning prism within a nitrogen-purged glass bubble, which protruded from the lower fuselage. There was no ejection seat, nor automatic canopy jettison. Pilots in trouble would have to open the canopy themselves before bailing out.

The CL-282's T-tail was replaced by a conventional tail section, to be constructed as a single assembly including the rear fuselage. This was attached to the forward fuselage by just three 5/8-inch tension bolts. The size of the wing meant placing the tail further aft, and the fuselage was lengthened to 50 feet. But Kelly's gust control innovation reduced the tail loads and allowed a lighter structure.

Bob Wiele did the detail wing design after Johnson specified the section, the limit load factor (still 2.5g), and the aspect ratio (10.67). Compared with the CL-282, the result was a 20% increase in wing area to 600 sq ft and a 10-foot increase in wingspan to 80 feet. The drawing reached the edge of the boards before the wingtips could be added![8] The wing had a three-spar construction, but the conventional rib stiffeners were replaced by an unusual latticework of aluminum tubing.

The bigger wing allowed more fuel, which was carried in four integral wing tanks, feeding to a fuselage sump tank. The total of 1,335 gallons stretched the aircraft's endurance and range impressively, to ten hours and over 4,000 nautical miles. The pilot had a fuel quantity gauge for the sump tank only—no indication of how much fuel had been used from the wing tanks. He had to rely on a simple subtractive-type fuel counter which showed him how much was left (in gallons), backed up by a warning light which came on when only 50 gallons remained.

But which fuel to use? At high altitude, the combination of still-warm fuel and reduced pressure would cause the more volatile portions of standard jet kerosene to boil away through the tank vents. After seeking advice from NACA and Colonel Norm Appold at WADC, Johnson turned to General Jimmy Doolittle, who had served on the TCP steering committee and was a board member of the Shell Oil company. The famous wartime General got Shell to adapt a special low vapor pressure kerosene as a new jet fuel. It was designated LF-1A by Shell and JP-TS (for Thermally Stable) by the military.[9]

The plane did not have a name yet. It was Lockheed model number 185, but the engineers referred to it obliquely as the Article. However, Kelly Johnson began calling it the Angel, since he expected this unique design to soar so close to heaven!

A new pressure suit

Then there was the need to protect the pilot. The new aircraft would be cruising at least 20-30,000 feet above Armstrong's Line, at altitudes where body fluids would start to boil if cockpit pressure was lost. Bg Gen Don Flickenger, ARDC's expert in pilot protection, was summoned to Trevor Gardner's office in the Pentagon. Flickenger explained that the newly-developed partial pressure suits provided a "get me down" protection until the aircraft reached a physiologically safe altitude of about 10,000 feet. Gardner told him that get-me-down protection wasn't good enough for the type of operation that this plane would be flying! Flickenger reckoned it would take three to five years to improve the suits. Gardner gave him ten months. The General set to work with the Aeromedical Laboratory at Wright Field, the Lovelace Foundation at Albuquerque, and his favorite pressure suit contractor. This was the David Clark Company in Worcester, MA, which also manufactured some of the nation's finest brassieres and corsets! Flickenger told David Clark to produce a suit which could be worn for up to 12 hours uninflated, and for four hours if it inflated following a cockpit pressurization failure. Clark turned the detailed design work over to his pressure suit specialist, Joe Ruseckas.[10]

Engine troubles

That unforgiving high altitude regime was troubling Pratt & Whitney. They had to produce a powerplant which would be safely operated at the limit of EGT (Exhaust Gas Temperature) for not only takeoff and climb, but also for an eight or nine hour cruise. In early February, Wright Parkins told Kelly Johnson they wanted to abandon the higher thrust, lower-weight -31. Wouldn't the -37 suffice instead? A furious Johnson insisted they continue. He had accepted the -37 as an interim engine for early flight tests only. Johnson had assigned Elmer Gath to engine integration, with Ben Rich doing the inlet design. They were producing a near-perfect ram air distribution within the duct, to ensure the maximum delivery of thin upper air to the compressor face. P&W ploughed on with the -31. They decided to improve the oil seals, halve the oil tank size, and re-examine the oil cooler, alternator, and hydraulic pump. They would forge the turbine blades instead of casting them.[11]

Developing the sensors

Not far from Burbank in Pasadena, CA, a special team led by Bill McFadden began work on the new aircraft's cameras at Hycon—Trevor Gardner's old company. Land's group had recommended that they start by improving some standard USAF K-38 cameras which had 24-inch focal lengths and exposed 9 x 18-inch frames of film. These were improved by adding thermal stabilization, lightweight barrels, and pre-focused lenses. The film magazines were modified to accept up to 1,800 feet of Kodak's new mylar-base film. In a configuration designated A-1, one of these improved 24-inch cameras would be flown in a rocking mount for medium-scale coverage, alongside three 6-inch wide-angle mapping cameras in a standard trimetrogen arrangement. A second configuration designated A-2 consisted of three of the 24-inch cameras looking left and right at 37 degrees, and vertically down. Thanks to the new film, a complete A-2 rig with enough film for a 3,000-mile flight with stereo overlap, weighed only 339 lbs.[12]

The real innovation, though, would be the B-camera. By early 1955, Jim Baker had revised his earlier ideas in favor of a unique configuration which violated many of the current rules of the recce business. It combined the area coverage of a panoramic camera with the definition required for target analysis. Hitherto, only large fixed-focal length cameras had been able to provide the latter. A 36-inch focal length lens, together with the lens cone, shutter, and mirror, would rotate to any one of seven overlapping "stop-and-shoot" positions from horizon to horizon. The positions overlapped so that the photo-interpreters could perform stereo measurement. The lens aperture was set at f10 to ensure reliable pre-set focusing at very high altitude. A single swiveling mirror replaced the usual but much heavier prism. Two huge film magazines carried 9.5 x 18-inch film, which would wind from opposite directions and be simultaneously exposed to produce an 18 x 18 negative. The contra-winding idea was good news for Kelly Johnson; it meant that the camera's center of gravity remained unchanged through the flight, despite using up to 6,500 feet of film from each magazine. The B-camera weighed 577lbs when this amount of film was carried, but if only 4,000ft of film was loaded on each magazine, a reduction to 484lbs weight was achieved.[13]

Baker turned the concept over to Hycon for detailed design and engineering development. Meanwhile, he began work on the lenses for both A and B camera configurations under a subcontract from Perkin Elmer. Rod Scott of Perkin Elmer devised a 70mm tracking camera by scaling down a 48-inch panoramic camera design. The "tracker" would operate throughout the flight, and its comprehensive (though small-scale) coverage would help the photo-interpreters place the images provided by the 24- and 36-inch lenses in their geographic context.[14]

Baker pioneered the use of computer programs to calculate the optimum lens curvatures and spacings. He used Harvard University's IBM Current Program Calculator, but had to work on it overnight, in order to preserve secrecy. As a result, the Land group's ambitious goal for photographic resolution of 60 line-pairs per millimeter was often achieved once the B-camera went into service. From 60,000 feet, it could resolve objects on the ground as small as 75cm/2.5 feet across.

At the same time, Baker continued design work on the very long focal length, or C-camera. He argued with Kelly Johnson over the amount of space in the equipment bay which could be allocated to the C's 180-inch folding optics. The aircraft designer wanted more room for batteries and other support equipment. Development of the C-camera fell behind schedule, and it was eventually realized that the A and B camera systems were more than adequate to the task. (The prototype C-camera was turned over to the USAF, but was never used operationally).

Land and the TCP had recommended from the outset that the aircraft carry signals intelligence sensors as well as cameras. The task was allocated to Dr. Bert Miller at Ramo-Wooldridge. This was the Los Angeles corporation, recently founded by two former Hughes electronic engineers, which Trevor Gardner had nominated to project-manage the nation's urgent ICBM and satellite developments. (Gen Bennie Schriever, whose AFDAP shop had been instrumental in defining the new spyplane requirement, had transferred to the West Coast to take charge of this effort in mid-1954).

Ramo-Wooldridge faced the same type of challenge as Hycon: they had to radically reduce the size and weight of airborne sensing equipment, and make it run autonomously for up to nine hours. Miller came up with a miniaturized ELINT (electronic intelligence) system that would cover the S-, C-, and X-bands used by Soviet radars. It would determine PRF (Pulse Repetition Frequency) and scan rate, and fit inside the aircraft's nose. The tape recorder used a new type of thin, 1/4-inch tape that was subsequently used in the first mass-market tape recorders.

Secret Base
By late March 1955, the wind tunnel tests had been completed and fabrication was well underway, managed by Art Viereck. He used the big presses in Lockheed's main production factory at night, and removed the secret aircraft's parts to the Skunk Works in Building 82 before the dayshift arrived.

Johnson summoned Tony LeVier from Lockheed's new Palmdale facility, where he was working as chief test pilot for the XF-104. LeVier ageed to test-fly the Angel before Johnson would reveal the details. When he finally did so, LeVier declared that Kelly was switching him from flying the plane with the shortest wings in the world, to the one with the longest! Johnson then told LeVier and Skunk Works' logistics specialist Dorsey Kammerer to find a remote base where the new aircraft could be flown. The government didn't want it tested at Edwards or Palmdale, in full public gaze.

LeVier and Kammerer took the Skunk Works' Beech Bonanza, made a two-week aerial survey of the remote deserts of southern California and Nevada, and came up with a shortlist. None of them appealed to Bissell or Ritland. In mid-April, LeVier flew them and Johnson up to the most likely sites near the Nevada test range. Ritland was already familiar with the area, since he had flown all over it during his previous assignment with the B-29 test squadron, which dropped nuclear weapons there. He directed LeVier towards an old WWII airfield which he remembered, just north of the test range and next to the Groom Dry Lake. "We didn't even get a clearance, but flew over it, and within 30 seconds we knew it was *the* place," he said later. They landed on the lakebed. Johnson liked the site, but feared it would prove impossible to get it cleared with the Atomic Energy Commission (AEC).[15]

Ritland took quick care of that, and worked similar miracles from his one-man office in the Pentagon, whenever USAF roadblocks had to be cleared.[16] In June he was joined by reconnaissance specialist Col Russ Berg and Lt Col Leo Geary. The latter had previously been responsible for USAF support to the Agency's Air Maritime Division, and was therefore already familiar with the secret world. Over at DPS, meanwhile, Bissell appointed Jim Cunningham as the executive officer. He was a former Marine Corps pilot who had proved to be a smart administrator of various covert projects since joining the CIA. Cunningham and Geary would do the same jobs for the next ten years, providing valuable continuity and a wealth of experience.[17]

Such was the pace of the project, that paperwork always struggled to catch up. Lockheed had been working for two months before it sent a contract to the government! At one stage, Kelly Johnson took out a $3 million bank loan to cover production costs, when the CIA's payments fell behind schedule.[18] To preserve secrecy, the CIA's payment checks were made out to "The C & J Manufacturing Company" and sent to Kelly Johnson's home address. Subcontractors also dealt with this "front" company, making deliveries to an unmarked warehouse in downtown Burbank so that their work could not be associated with Lockheed.

There wasn't much paperwork, actually. The monthly progress reports were kept short; unlike regular military business, there were no teams of plant representatives sent in to second-guess the contractors. Bissell told Johnson: "When you face a decision, I want to know about it and pass judgement. But I rely on you to tell me the potential consequences and costs." In this atmosphere of trust and dedication, the program moved smartly ahead. As it transpired, Lockheed underspent on the first contract worth $22.5 million, which covered the 20 aircraft for the CIA. The corporation returned $2 million to the government, meaning that these airframes cost only just over $1 million each!

In early May, Herb Miller issued contracts from DPS worth $0.8 million for construction of the secret base. Johnson had promised to fly the prototype in August, and doubted whether the base would be ready by then. Washington had turned Kammerer's plan for a small, temporary facility into a larger, permanent facility with a runway, control tower, mess hall, and three hangars. But they were starting from scratch in that desert wasteland: no roads, no water, no power. With some irony, Johnson named the place Paradise Ranch.[19]

OILSTONE and AQUATONE

By the third week in May, the prototype "Article" was out of the jigs. But the wing was way behind schedule; aileron design was not yet finished. Johnson put almost everybody on it, and worried about bending instability.[20] He appointed Ernie Joiner to run the flight test program, and sent his contract for Lockheed's role in the subsequent training and deployment phases to the CIA. Cunningham arrived at Burbank to begin the strict security checks on those engineers who had volunteered to go on the deployments.

By late June, it became apparent that the respective roles of the CIA and the USAF in this enterprise needed formal definition. Bissell and Ritland flew to SAC headquarters to brief General LeMay. When LeMay realized how much the regular USAF had been cut out of the loop, he was furious. So was General Thomas Power, the commander of ARDC, when Gardner and Putt briefed him into the project. LeMay and Power enlisted the support of USAF Chief of Staff Nathan Twining, and forced a showdown. The meeting was held in Colorado Springs in early July 1955. CIA Director Dulles and Secretary of the Air Force Harold Talbott rejected LeMay's bid to run the operational phase, and reaffirmed the basic division of responsibilities. But the powerful general insisted that SAC help select the pilots, run the training program, and staff some key operational slots within DPS.[21] The codename for USAF support of the program was changed from "SHOEHORN" to "OILSTONE." The CIA had adopted the cryptonym "AQUATONE." The secret aircraft still didn't have a proper name or designation...

Johnson returned from a short vacation to lead a terrific drive to finish the airplane. Building 82 was working round-the-clock. In early July, the static test airframe was ready. Henry Combs, Dick Hruda, and Ray McHenry had done the stress analysis, fuselage landing gear, and wing, respectively. Had they set the correct limit load requirements, or had too much been sacrificed in the desperate quest for altitude? John Henning managed the static tests, and the critical load limits were proved by the end of the month. The horizontal tail had to be beefed up a little, but the basic design passed the test.[22]

Between 17 and 21 July, Johnson made a detailed inspection of the prototype. Three days later, Ritland arrived from Washington to witness loading of the now-disassembled article into a C-124 transport, which would fly it to Paradise Ranch.

As part of the contract, Lockheed had designed two wheeled ground carts so that the Article could be easily airfreighted. One carried the aft fuselage, wings, and tail surfaces. The other was adapted from the original CL-282 take-off dolly design, and it carried the forward fuselage complete with the installed engine. Much of the airplane maintenance could be carried out while it was on these carts, which had self-contained hydraulic and electrical systems.

Incredibly, the secret base in the Nevada desert was at least semi-complete. Herb Miller had organized a large team of construction crews via the Atomic Energy Commission, and they had also worked round-the-clock. The tarmac runway had just been laid. Newly-appointed base commander Dick Newton clashed with Johnson over whether it was ready to accept the heavy C-124. As usual, Johnson won the argument. The Article was flown in, unloaded, wheeled into a semi-complete hangar, and re-assembled.[23]

Soon, the Article was ready for engine runs. But the J57 wouldn't start with the special fuel. The ground crew found some five-gallon drums containing JP-4, linked them together, and ran a hose from them into a fuel valve in the J57. Then they lit the engine with the conventional fuel before withdrawing the hose! The proper fix to the starting problem was soon devised—make the engine spark plugs longer—but the interim solution was typical of the "can do" approach which moved the program forward at such a rapid pace.[24]

The Angel Soars

The first taxi trials took place on the lakebed on 1 August 1955. The second run turned into a first flight of sorts when, much to LeVier's surprise since he had already chopped the power to idle, the aircraft left the ground at around 70 knots and flew for about a quarter of a mile with the flight test crew in hot pursuit. This bird's wing had all the lift expected...and more! It eventually made heavy contact with the ground in a 10-degree left bank. The Article bounced back up, then settled down again under semi-control and eventually skidded to a halt a mile further on. Both main tires had blown, and the brakes caught fire.

After more taxi tests, the Article was prepared for a proper first flight on 4 August. The day dawned unsettled, with thunderstorms threatening, but Johnson was determined to press on, so in mid-afternoon they wheeled her out again. At 3:55 pm the strange-looking plane with a polished metal finish soared into the sky. "It flies like a baby buggy!" exclaimed LeVier to fellow test pilot Bob Matye, who was flying chase with a cameraman in a T-33. Kelly Johnson and Henry Combs followed at a safe distance in a C-47, piloted by another Lockheed test pilot, Ray Goudey.

LeVier was less enthusiastic after he brought the Article in to land after an uneventful twenty-minute flight to check basic handling qualities. He had previously debated the landing technique with Johnson, who insisted that his Angel be landed tail up, with the main gear touching first, to avoid stalling the wing. The test pilot reckoned it should be brought in to a two-point landing—"just like any taildragger," he said. Johnson prevailed, but on his first landing attempt LeVier ran into exactly the problem he had predicted. Despite partial flaps, fuselage speed brakes extended, and almost idle power, the aircraft had the flattest of glide angles, and when he attempted to grease the main gear at about 90 knots, the porpoising started.

LeVier applied power and went around, only to encounter the same problem again. And again! By now, the thunderstorms were closing in, and Johnson was getting very anxious. He even contemplated a belly-landing, but LeVier saved the day by trying his taildragger theory, holding the nose up into a flare and executing a perfect two-pointer just as the aircraft neared the stall. The pogos had been locked in position for this first flight, and even at 60 knots with the gear on the runway they were still airborne. The aircraft started to porpoise again. LeVier managed to control it, and the Article finally rolled to a halt at 4:35 pm. Ten minutes later, a downpour flooded the lake with two inches of water! This didn't dampen the spirits of the Skunk Works team, who spent the evening celebrating with beer-drinking and arm-wrestling contests.[25]

LeVier flew again on 6 August and practiced his landing technique. On 8 August at 9 am, the Article made its *official* first flight in front of Bissell, Ritland, and a few more government representatives. It was a spectacular sight, for the power-weight ratio was so great that the take-off roll was just a few hundred feet. The aircraft had to be pulled into a steep climb as soon as it left the ground, to keep below limit speed. After one pass over the spectators, LeVier went on climbing to 32,000 feet as planned. Matye and Johnson struggled unsuccessfully in the T-33 to catch up. The hour-long flight ended with another low pass and a reasonable landing. Kelly had kept his promise, and flown the Angel within just eight months!

Flight Test Problems

An anti-porpoise valve was added to the main gear to help landings, and the flight tests resumed. LeVier expanded the flight envelope at low-medium altitude, and checked out Bob Matye and Ray Goudey in the Angel before being reassigned at the end of August.[26] By then, Matye and Goudey were making one flight nearly every day, and the Angel had soared past 60,000 feet. But the flameouts had begun.

The interim -37 version of the J57 engine was mainly to blame. The fuel control was inadequate, and the bleed valves didn't work properly. One false move with the throttle and the engine would quit. The compressor stalls would occur from 45,000 feet and upwards with a loud banging noise, which startled the pilots. They experienced flameouts under a variety of conditions; during a slight yaw or speed reduction, for instance. At first the throttle could not be retarded to idle at all without the engine flaming out. Since the engine could not be restarted in the thin upper air, the aircraft had to descend to 35,000 feet or lower in order to get a relight. At high altitude, the -37 also leaked oil through the high-pressure compressor that ran the cockpit airconditioning. An oily film would accumulate on the inside of the cockpit canopy. The pilots were obliged to take along a swab mounted on the end of a stick, in order to wipe the windshield clean! Since cockpit air was also used to pressurize the equipment bay, the oily film was also deposited on the camera windows, making photography useless.

Skunk Works engineer Ben Rich came up with a temporary solution: stuff sanitary napkins around the oil filter! Then the bearing seals and oil scavenge system were improved, but the problem was never completely solved in the -37 engine.[27]

The early flight tests also revealed fuel feed problems. The wing had been designed with fore-and-aft fuel tanks, rather than inboard and outboard. In a steep climb, the aft (auxiliary) tanks would not gravity-feed the sump tank. Pressurization had to be added, and also a ram air scoop to equalize tank pressure in a rapid descent. Uneven fuel feed was another problem, solved eventually by adding a small electric pump.

Flights were restricted to a 200-mile radius of the Ranch. The article could glide at least that far from high altitude and make a deadstick landing, if a relight wasn't possible!

The test pilots developed some standard flight profiles. The aircraft would be climbed rapidly to 55,000 feet, from where a specified speed schedule would be flown. After a somewhat slower climb over the next 10,000 feet, the aircraft would enter a "cruise-climb" for the remainder of the flight, climbing slowly higher as the fuel burnt off. The maximum altitude it might reach depended on a number of variables, such as aircraft take-off weight and outside air temperature. But as it crept higher, the precious margin between the aircraft's stalling speed and its maximum speed eroded quickly, so that by about 68,000 feet in an aircraft that had taken off fully-fueled, there could be as little as ten knots difference between stall buffet and Mach buffet.

In other words, the slowest the aircraft should be flown was approaching the fastest it should be flown! This was "coffin corner," a condition that afflicted all high-altitude flyers, but in the Angel the condition was particularly acute. And it was difficult to tell the difference between stall buffet and Mach buffet. Recovery from a stall could be difficult, since control surface response was poor in the thin upper atmosphere. If the limiting Mach number of 0.8 was exceeded by much, the nose would pitch down and the aircraft would begin accelerating. If the pilot couldn't regain control, it might eventually break up at lower altitude, where the placard speed was only 260 knots, even with the gust control deployed.

Organizing, Growing

The formal OILSTONE-AQUATONE agreement was signed by Dulles and a reluctant Twining on 3 August 1955. It specified that Bissell should have a military deputy with responsibility for deployment and operations. In late August, Ritland moved from the Pentagon to project HQ and took up this role. By now, DPS had moved out of the CIA's administration building to another address on E Street. Soon, it would move again to "Quarters Eye" on Ohio Street NW, before finally settling into the fifth floor of a downtown office block in 1956. This physical separation from the rest of the CIA added to the strict compartmenting of the project for security purposes—and to the mystique which grew to surround it. The project even had its own secure communications set-up. CIA staffers like John Parangosky and John McMahon, who started their CIA careers in DPS as an administration and security officer, respectively, gained the experience and kudos here which would stand them in good stead later in their careers.

In Washington, Kelly Johnson lobbied for more airplanes. He proposed fitting the Westinghouse APQ-56 side-looking reconnaissance radar, already in production for SAC's RB-47Es. He suggested an "intruder version" of the aircraft armed with two small nuclear bombs, and actually got Gardner's permission to start design of a "long range interceptor" version equipped with a Gatling gun in the Q-bay.[28] As more top commanders were allowed knowledge of the program and its amazing progress, the list of Lockheed fans grew longer. Like LeMay and Powell before them, though, many were upset about previously being denied knowledge of the project. In mid-September, General Al Boyd at Wright Field was finally let in on the secret. The WADC commander insisted that he be allowed to fly the plane—immediately!

It was already obvious that Lockheed had built a winner. Prompted by Gardner, Assistant Secretary of Defense Donald Quarles agreed to buy more aircraft, this time for the USAF. In early October, the axe fell on the rival Bell X-16 program. It came as a complete shock to Dick Smith and his design team in Buffalo, NY. They were ahead of schedule, the first flight was due early in 1956, and a contract for 22 production aircraft had just been received from ARDC. They knew nothing about the "Lockheed super-glider," and the cancellation of the X-16 was quite a blow for a relatively small company.

In early November, Johnson went back to Washington with a recommendation that the interceptor version be put on hold. It was never built, but Johnson returned to Burbank with a provisional agreement to build 29 more of the basic reconnaissance version, this time for the Strategic Air Command (SAC). Leo Geary reckoned it was time to give the Angel a proper military designation. But its true purpose still could not be revealed. Geary called Colonel Jewell Maxwell at ARDC, who dealt with aircraft designations. Geary and Maxwell agreed to put the new aircraft into the U-for-utility category. The first number had already been allocated, so the Angel became the U-2.[29]

Pilot Selection begins

The pressure to deploy remained as great as ever. Critics of the Eisenhower administration said that a "bomber gap" was opening up, since the Soviets seemed to be accelerating development and deployment of their nuclear strike force. There was still no sure way of knowing. At a summit conference in Geneva in July 1955, new Soviet leader Nikita Khrushchev had given a frosty reception to President Eisenhower's "Open Skies" proposal for the USA and the USSR to conduct reciprocal aerial reconnaissance flights over each other's territory. In November 1955 the USSR formally rejected the Open Skies proposal. The U-2 pilot training program had to begin, although flight tests were far from complete.

The CIA's Air Maritime Division had been running covert air operations behind the Iron Curtain since the late forties, to insert agents and saboteurs, and drop propaganda leaflets. Defectors and stateless nationals from Eastern Europe were taught to fly these missions under Project OSTIARY. Someone suggested that the "Ostiaries" be trained as U-2 pilots for the overflight mission. If one was shot down, it was argued, the U.S. could deny direct involvement. CIA Deputy Director Pierre Cabell didn't like the idea.[30] Leo Geary, who had supported the CIA's covert flights during a tour at the MAAG in Athens from 1950-53, instead suggested some Greek pilots who had flown specially-modified P-51s over Albania and Bulgaria. He arranged for them to enter the USAF jet training course at Williams AFB, AZ.[31]

Meanwhile, General LeMay selected Colonel Bill Yancey to organize a pilot training program at the Ranch. Like Colonel Marion "Hack" Mixson, whom LeMay had already installed at project HQ to run the deployments, Yancey was an old SAC recce hand. Lt Col Phil Robertson had succeeded him as SAC's reconnaissance branch chief, so Yancey selected him to be operations officer. Yancey and Robertson chose four more pilots from varying backgrounds: Lou Garvin and Bob Mullen from a B-47 wing because they had "taildragger" experience; plus Hank Meierdierck and Lou Setter, who had both been flying jet fighters for SAC.[32] They arrived at Burbank in early November, and the Skunk Works put them through a rudimentary ground school. At the same time, General Boyd and renowned USAF test pilot Colonel Pete Everest made their flight evaluation of the Article at the Ranch. They gave it ARDC's formal seal of approval, though Johnson noted in his diary that "Everest couldn't land the plane."

Pressing on with Flight Tests

Lockheed test pilot Bob Sieker had joined the program in early September 1955, replacing LeVier. Goudey, Matye, and Sieker flew intensively, but there were still only two Articles at the Ranch and many problems to resolve. Ernie Joiner was running flight tests with the help of just four engineers (Paul Deal, Glen Fulkerson, Bob Klinger, and Mitch Yoshii). Dorsey Kammerer managed the maintenance, modification, and supply with only another 20 Lockheed men. A fifth test pilot, Bob Schumacher, arrived in mid-November.

The -37 engines still coughed and "chugged" disconcertingly at higher altitude. There were turbine blade failures. More flameouts, more relights, sometimes three per flight. The autopilots were late being delivered from the Lear company. Without them, the aircraft needed constant attention in the cruise-climb. The test pilots managed this, but operational pilots could hardly be expected to monitor the driftsight, take navigation fixes, operate the camera system, and so on at the same time. When the autopilots did finally arrive, the trim had to be frequently reset to fly the cruise-climb airspeed schedule, which was not a constant Mach number. The test pilots encountered sudden temperature changes at the tropopause, which could cause the autopilot to over-react and input too much pitch. They called this phase of flight "the badlands."

Despite these problems, the Angel soared to its design altitude of 70,000 feet, well past the published world record of 65,890 feet. This had recently been set with some fanfare in the UK by Wing Commander Walter Gibb flying a specially-modified Canberra. There were no triumphant public announcements from the Nevada desert, though!

The prototype Article had strain gages installed in the same locations as the static test article at Burbank. Structural flight tests were initially conducted with the aircraft restricted to 80% of design limit load. Then the measured flight data was correlated with the static test data, before the static tests were extended to ultimate load. Ultimate loads of some 4g were measured, but the limit loads were kept at 2.5g, as originally conceived. The limit loads were duly demonstrated in test flights. The U-2's structure was therefore proved in an orderly fashion, despite the pressure of time. Moreover, the airframe proved stronger than many had thought.

To cater for the growing number of commuters, the USAF started a daily shuttle service to the Ranch from March AFB and Burbank with a C-54 transport. At 0700 on the morning of Wednesday, 17 November 1955, it departed Burbank as usual with 14 project personnel onboard, plus five crew. In poor weather, the C-54 smashed into the top of Mount Charleston near Las Vegas, killing them all. If it had been only 30 feet higher, the lumbering transport would have cleared the 11,000 foot peak. It took three days for rescue parties to reach the wreckage; they were accompanied by a USAF colonel who removed briefcases and top-secret U-2 documents from the charred bodies of the victims. The dead included project security chief Bill Marr and four of his staff from the CIA; Dick Hruda and Rod Kreimendahl from the Skunk Works design team; and Hycon's vice-president, H.C. Silent. In an official statement, they were described as civilian technicians and consultants, and the secrecy of Project AQUATONE was preserved as outsiders concluded that the deceased were all working for the Atomic Energy Commission on nuclear weapons projects at the Nevada test range. By sheer chance, some members of the Skunk Works flight test crew missed the flight. They had hangovers from a party the previous night. The USAF shuttle was soon resumed, but Kelly insisted that his own people fly the Skunk Works C-47 to and from the Ranch.

Recruitment

Lockheed's test pilots taught Yancey's training group how to fly the plane. This was inevitably a haphazard affair; most of the U-2's unique flying characteristics could only be experienced first-hand on the initial solo flight. The instructor pilot flew alongside in a T-33 and shouted instructions and encouragement by radio. Landings were still a big challenge: approaching the stall, each aircraft behaved a little differently. Some would roll-off in one direction, some in the other. The flight test engineers added small fixed strips to the leading inboard edge of the wing, tailored to each aircraft's individuality as it approached the stall.

Four Greek pilots arrived—the others had been washed out at the training base. It quickly became obvious to the SAC training group that they could not fly the U-2. One nearly wrecked an aircraft by landing it wing-low. Lt Col Phil Robertson recommended that the project use pilots from SAC's F-84 squadrons instead. Some of these men had combat experience from the Korean War, and had flown long overwater deployments in the single-engined fighter. Flying over hostile territory in the U-2 wouldn't be so different, he reckoned.[33] Conveniently, SAC was disbanding its four Strategic Fighter Wings which flew these F-84s at Bergstrom AFB, TX, and Turner AFB, GA. The security staff at project HQ liked the idea. The SAC pilots were all facing re-assignment, so if some of them disappeared to a secret project, it might not be noticed.

Leo Geary, Hack Mixson, and Jim Cunningham began screening the personnel records for suitable reserve officers. The best candidates were carefully approached, and turned over to the Agency for exhaustive interviews, medicals, and background checks. They wondered at the intense security; initial screening interviews in an off-base motel, false surnames and ID cards, and lie-detector tests.[34] But the money was good: $2,500 a month. This was triple their military pay, although the CIA would retain $1,000 per month until each pilot had completed his two-year contract. The selected pilots were sent to Wright Field for an altitude chamber test, a ride on the centrifuge, and other stress tests. From there to the Lovelace Clinic at Albuquerque, where they underwent four days of most intrusive medicals, supervised by Dr Donald Kilgore.[35] These prospective U-2 pilots were effectively the guinea pigs for the intense physical and psychological screening which would later be made famous by the Project Mercury astronauts. "Barium was inserted everywhere you could think of," noted Bill Hall, one of the first U-2 recruits. "We all managed to pass the course, but I thought that breathing into air bags while riding a stationary bicycle would do me in!"

Don Flickenger had set these demanding physiological requirements, for which he would no doubt have been roundly cursed by the pilots, had they known. But when Flickenger left ARDC in late 1955 for a new post in Europe, his legacy to Project AQUATONE was a radically improved life support system compared with anything else then flying. The USAF's top flight surgeon left Colonel Phil Maher of the Aeromedical Lab to complete its development with the two key contractors. At the David Clark Company, Joe Ruseckas finished the design of the MC-3 partial pressure suit. Compared with the old T-1 suit they had been using, pilots found the new suit more comfortable, with a pressure bladder and extra sets of adjustable lacing across the chest. They no longer had to "reverse breathe," that is, exhale forcibly in order to take in oxygen. In the early years, each suit was custom-tailored, and the pilots traveled to the Clark factory at Worcester to be fitted. Meanwhile, the Firewel division of Munsingwear Industries in Buffalo, NY, provided the oxygen system, regulators, and pilot's seat kit.

The interim life support equipment had already saved the early U-2 pilots' lives more than 20 times after high-altitude flameouts. Before each high-altitude flight, pilots were required to "pre-breathe" oxygen for two hours to eliminate nitrogen bubbles from the bloodstream. Otherwise, they would experience pain equivalent to the "bends" experienced by divers during a decompression. Later, it was realized that individuals vary in the rate at which they eliminate nitrogen, and one hour of prebreathing was sufficient for most pilots. A flight surgeon was always on hand to monitor preflight preparations.

As the test flights were extended in duration, it became apparent that the rate of oxygen consumption also differed considerably between pilots. Lou Setter from the SAC training group added an oxygen plot to the fuel plot he had already devised. Using these plots, pilots could monitor the rate of consumption of both fuel and oxygen, as the flight progressed. On one early flight to high altitude, Phil Robertson noticed from the plots that his oxygen supply was fast-depleting. He quickly initiated a descent, but by 40,000 feet he had only the emergency supply left. By 25,000 feet that, too, was exhausted, and Robertson had to open his helmet faceplate. He landed safely, and the loss of oxygen was traced to a joint in the supply line which deformed and leaked when the pilot sat on the seat pack. Development of the life support system, like most other aspects of the U-2, often proceeded by trial-and-error![36]

Refinement

Six SAC fighter pilots made it through the selection process to form the CIA's first group of U-2 flyers. They were obliged to resign their USAF commissions, but were promised a no-questions-asked return to the military with no break of service, upon completion of their contract with the CIA. The procedure was known as "sheep-dipping." As a cover story, the recruits became test pilots with Lockheed, but the CIA really paid their wages.

This first group began training at the Ranch on 11 January 1956. Yancey's men had devised a ground school and flight check-out syllabus. They took the newcomers for three flights in the T-33 to simulate high-altitude flameouts and restarts, and to demonstrate near-stall landings. Then came the first solo in the U-2, a flight to 20,000 feet followed by five practice landings on the lakebed. The awkward pressure suit was worn for the first time on the third flight, lasting three hours and reaching 60,000 feet. A further nine flights progressively introduced high altitude navigation and photography; night flying; eight-hour-plus missions; and runway landings. Having logged 58 hours in the U-2 and flown a final eight-hour check-ride, the new pilot was declared mission-qualified.

The Angel began to venture far from the secret base on endurance-proving flights, now that the test pilots had devised satisfactory climb and cruise speed schedules which took account of the temperamental -37 engine. On 2 February 1956, Article 344 flew for nine hours and 35 minutes, the longest-yet. Bob Schumacher had taken off with a maximum fuel load of 1,362 gallons at a gross weight of 19,900 lbs. He had flown until the fuel low-level warning light came on. But the simple, subtractive fuel counter (which was the only fuel quantity information provided in the cockpit) showed that 1,386 gallons had been consumed, 24 more than was loaded! Furthermore, another 15 gallons was drained from the aircraft after landing.

Where had the extra 39 gallons come from? It was discovered that the U-2 had a unique ability to "manufacture" fuel. The kerosene from the wing tanks which had been "cold-soaking" for hours was heated—and therefore expanded—as it passed through the fuel-oil heat exchanger! However, for operational flight-planning purposes, it was decided to stick to the original design maximum fuel capacity of 1,335 gallons and gross weight of 17,300 lbs. When fitted with the J57-P-37 engine, the U-2 therefore had a maximum range of about 3,500 nautical miles, and an endurance of eight and a half hours. The -31 engine promised to give the U-2 better range and altitude—when it eventually arrived.

The long-range flight tests in early 1956 also served to develop the aircraft's navigation system. The basic navaid was the driftsight, the "downward-looking periscope," which provided a 360-degree view beneath the aircraft at two selectable magnifications. But what if it was cloudy below? Dead-reckoning was a viable alternative, since the winds were usually quite light at high altitude. There was also a radio-compass, the AN/ARN-6. But radio stations might be few and far between over denied territory. These three basic means of navigation could not provide the accuracy, reliability, and redundancy which would be required for such long flights.

The known system of long-range, high-frequency ground stations offered a possible alternative, since the pilot could tune into these and determine his position by triangulation. An HF receiver and antenna unit from the Collins Radio company was selected for the U-2 and initially tested successfully in a C-47. But when it was added to a U-2, there were serious problems with the long-wire antenna installation. When the CIA realized that the use of HF could betray the aircraft's position to enemy air defenses, the device (which was designated System II) was dropped.

In late 1955 it was finally decided to adopt celestial navigation. Dr Robert Hills of the Baird Atomic company in Boston quickly designed a sextant which could be coupled with the driftsight optics, using a levered mirror. Above 50,000 feet, the sky was black from overhead to near the horizon even at midday, allowing star fixes to be taken. The pressure-suited pilot couldn't perform the necessary calculations in the air, however, so a flight-rated navigator precomputed them before takeoff. Major Ray Burroughs from SAC HQ and Jack DeLapp in the SAC training group at the Ranch, refined these techniques. Like the many U-2 navigators who followed, their preflight work on the ground was vital to the success of the mission.[37]

The prototype of the improved J57-P-31 engine arrived at the Ranch in January 1956 and was fitted to Article 341. Hand-built by Pratt & Whitney engineers, it weighed 450 lbs less and produced 700 lbs more thrust than the 10,500 lb thrust of the -37. At 70,000 feet and above, it burnt just 700 lbs of fuel per hour. And it didn't leak oil, or flameout so readily! Pilots found the -31 much easier to operate, since only one variable (EGT) had to be monitored. However, there was a downside to the new engine: the air restart characteristics were not as good as the -37 because of the lower minimum fuel flow. The -31 couldn't be restarted above 35,000 feet at all.

Ready to Deploy!
Lockheed's pilots completed the basic flight test program at the end of February, although proving of the new engine continued. Remarkably few changes to the basic airframe were required. Although the B-camera was not yet ready, the A-camera configurations were flown, and the aircraft proved to be a very stable reconnaissance platform. More milestones were reached. By late March, a maximum altitude of 73,800 feet had been demonstrated using the new engine. Maximum range was around 4,000 nautical miles, and endurance was nine hours 30 minutes. Total flight time reached 1,000 hours. Nine airplanes had been delivered.[38]

Incredibly, there had been only one serious accident. On 21 March, the second Article was damaged when CIA pilot Carmine Vito touched down on the lakebed in a yaw, while trying to correct for gusting winds. The main gear collapsed and punctured the sump tank.

There had been plenty of potentially serious incidents, though. The engineers were still fiddling with the leading edge, applying layers of duck tape to try and improve the stall characteristics. CIA pilot Marty Knutson was checking this out inflight when the aircraft stalled and entered a spin at 20,000 feet. This had never happened before, and the anxious pilot from the SAC training group who was flying chase in a T-33 urged Knutson to bailout. He would have done so, had he been physically able to jettison the canopy! Instead, by using rudder against the spin and with the huge control yoke fully forward, Knutson recovered control and landed safely. A structural check revealed some damage in the wing, where fuel had been forced outboard during the spin by centrifugal force. This aircraft wasn't cleared for spinning![39]

The CIA was arranging a base for the first overseas deployment. The flyaway spares kits had already been designed. The USAF scheduled an operational readiness inspection at the Ranch. Planes, payloads, procedures, and operational pilots were all evaluated in a six-day "Unit Simulated Combat Maneuvers" (USCM) test from 9-14 April 1956. It went very well, although pilot Jake Kratt was forced to make the first landing away from base during his long-range navigation test. He flamed out over the Mississippi River, relit, flamed out again 600 miles further on, and stretched a glide all the way into Kirtland AFB, Albuquerque, NM.[40] Marty Knutson also had a flameout during his long-range mission, although he managed to relight and recover into the Ranch. He was flying a -37 engine aircraft, and Pratt & Whitney tech reps discovered that some of the turbine blades had failed. Prolonged operation at high EGT and high altitude had caused them to crystallize.

Col Yancey reported to General LeMay that the first group was ready for operations. Bissell and others had their doubts, especially since there weren't yet enough -31 engines to fit the deployable aircraft. A compromise was reached: the group would deploy, and the more reliable -31 engines would be shipped to them direct. Until they were installed, no overflights of denied territory would be authorized. In the first days of May, four planes and their support gear were loaded into C-124s, which took off. As usual, only those who "needed to know" were aware of the destination.

Notes:

[1] Ritland papers, in AFFTC History Office, Edwards AFB

[2] see Appendix

[3] Richard Bissell, "Reflections of a Cold War Warrior", Yale University Press, 1996, p92-. Bissell's memoirs were completed after his death by his personal assistant in later life, Frances Pudlo, and researcher Jonathan Lewis. Hereafter Bissell memoirs. Readers will find significant differences between this author's account of events, and those as related by the Bissell memoir. It should be noted that neither Bissell, nor his co-authors, had access to classified or declassified documents.

[4] Johnson Log, 15-17 December 1954

[5] Henry Combs interview by author, and remarks at "The U-2: A Revolution in Intelligence" conference, Washington D.C, September 1998, hereafter "U-2 conference"

[6] patent filing on "Droppable Stabilizing Gear for Aircraft", dated 6 September 1955, provided by Lockheed Martin Skunk Works

[7] Ed Baldwin interviewed by author

[8] Baldwin interview

[9] Sloop NASA history, p127. During the flight test program, it was determined that standard JP-4 could be used in the U-2 at low-medium altitudes for ferry purposes, with a restricted rate of climb, but not above 50,000 feet.

[10] letter from Flickenger to the David Clark Company, 1987, reproduced in the company's David Clark memoirs, 1992

[11] Johnson log, February 1955. Ben Koziol interviewed by author. Ben Rich "Skunk Works", Little Brown and Co, 1994

[12] soundtrack to the film "The Inquisitive Angel", made by the CIA in 1957 and obtained by the author

[13] Jim Baker interviewed by author, and U-2 conference, plus technical manual SP-0045 for the B-camera

[14] Formally, Perkin-Elmer (P-E) itself also worked under subcontract to Hycon. Effectively, Baker did the most important work on the early U-2 lenses, including the final grounding, and P-E ordered the glass from by the German company Schott, per Baker's recommendation

[15] Johnson log, 12-13 April 1955; Maj Gen Osmand Ritland, USAF oral history interview K239.0512-722. The author prefers the evidence of these two sources to that of Tony LeVier, who claimed that he was responsible for 'discovering' Groom Lake.

[16] The Project OILSTONE office in the Pentagon was disguised as AFCIG-5, "Special Assistant to the Inspector General for Special Projects"

[17] Leo Geary interviewed by author.

[18] Ben Rich and Leo Janos, "Skunk Works", Little, Brown & Co, US 1994

[19] Johnson diary, May 1955

[20] Baldwin interview

[21] Bissell memoirs and "Origins of the U-2" article-cum-interview with Bissell in Air Power History, winter 1989, hereafter Bissell article; Ritland papers; Geary interview

[22] Combs, Baldwin, interviews

[23] Johnson diary

[24] Bob Murphy interview

[25] flight test reports as reprinted in "Secret First Flight of Article 001", Spyplanes, vol 2, 1988. However, for much of the subsequent technical details in this chapter on flight testing, the author has used Lockheed Report No SP-109, "Flight Test Development of the Lockheed U-2 Airplane", 4 November 1958, declassified and made available to the author, 1999.

[26] contrary to other accounts, LeVier did not fly the U-2 to high altitude

[27] Rich/Janos book, p138

[28] Johnson diary entries for 6 and 22-26 August; Baldwin interview; drawings of a design carrying small nuclear bombs on underwing pylons, discovered in Lockheed ADP archives by the late Richard Abrams

[29] Johnson diary, October/November; Geary interview

[30] author interview with anonymous source; Project OSTIARY has never been declassified

[31] Geary interview

[32] Hank Meierdierck, Phil Robertson and Lou Setter interviews by author. This group were officially part of the 4070th Support Wing, which SAC formed at March AFB to provide logistics support to the early U-2 program.

[33] Robertson interview

[34] Meierdierck, Robertson, Geary interviews

[35] Kilgore was at The Ranch for the early test flights, where he provided medical supervision for the Lockheed test pilots. "Lovelace Doctor Had Secret Role," Albuquerque Journal, 2 August 1987, pA1.

[36] Lou Setter, Phil Robertson interviews

[37] Johnson log, 1 December 1955; soundtrack to "The Invisible Angel"; Jack DeLapp interviewed by author

[38] Col Phil Robertson subsequently claimed an unofficial world altitude record of 74,500 feet during a 10 hour 5 minute flight on 10 May. He flew all the way north to the Great Bear Lake on Canada's Arctic Circle, in a flight designed to measure the effect of high-altitude wind patterns. Returning the way he'd come in a straight line, he arrived overhead The Ranch with plenty of fuel left. So he flew on to San Diego and back. The record was not beaten (again unofficially) until the re-engined U-2C flew in 1959. Robertson interview. (Six days after Robertson's flight, test pilot Bob Schumacher flew for 10 hours 15 minutes, which is the longest-recorded flight of a U-2A model).

[39] Marty Knutson interview

[40] Pedlow/Welzenbach p78-79.

3

Doing the Deed

During a visit to Washington in January 1956, the British Foreign Secretary, Selwyn Lloyd, met with Allen Dulles, Director of Central Intelligence. Dulles outlined Project AQUATONE to his British visitor, and asked for permission to base the first U-2 detachment in the UK. It was an obvious choice; the Anglo-American alliance was particularly strong in intelligence-sharing. The Burns-Templar agreement of 1950 had formalized the exchange and protection of all military and most intelligence information between the two countries.[1] The separate and even more secret UKUSA agreement covering Signals Intelligence (SIGINT) had been signed in 1948.

Selwyn Lloyd returned to the UK and recommended the project to Prime Minister (PM) Sir Anthony Eden. "Allen Dulles promised me that all the intelligence resulting from the operation would be shared with the British authorities," he wrote. "This would be greatly to our advantage, since Soviet airfields which are important to us would be covered."[2]

Eden was dubious. A political flap had just occurred thanks to a USAF attempt to overfly the Iron Curtain countries at high altitude with camera-carrying balloons. After five years of development and despite opposition from CIA and skepticism from the USAF's own engineers at Wright Field, the project had gone ahead. President Eisenhower approved the operational phase in late December 1955. Over 500 balloons were launched from Western Europe, the theory being that the prevailing jetstreams would carry them west-to-east across the Soviet landmass, so that they could be recovered 8-10 days later in northern Asia.

The scheme was a dismal failure, with less than 10 percent of the camera payloads recovered. Many others were shot down by Warsaw Pact fighters, or simply drifted down to earth. The USSR collected the evidence, displayed it in Moscow, and issued strong protests. The British PM wasn't impressed. He told aides that he wouldn't approve the proposed U-2 flights for the time being, "in view of the notoriety that the balloons have gained."[3]

But Bissell and Dulles applied pressure via Eisenhower, and Eden reluctantly changed his mind. On 1 May 1956 the first of four U-2s arrived at the USAF's Lakenheath base in East Anglia inside a C-124. By mid-May, they were reassembled and flying. Bissell and Kelly Johnson flew to London to supervise activities. Royal Air Force fighters were tasked to attempt practice intercepts against the U-2s—they didn't get near.[4]

Cover Story

An elaborate cover story had been devised to mask the true purpose of Project AQUATONE. On 7 May a press release was issued in the name of Dr Hugh Dryden, director of the National Advisory Committee for Aeronautics (NACA)—the predecessor of NASA. It announced that the first U-2 aircraft were flying from Watertown Strip in Southern Nevada in a new research program. Capable of reaching 55,000 feet, the aircraft would gather research data about the upper atmosphere to help plan future jet airliners which would routinely be flying at the higher levels. Jetstreams, clear air turbulence, cosmic ray particles, the ozone layer, convective clouds—all would be studied by the new aircraft, a few of which had been "made available" to NACA by the USAF. More detailed releases followed, giving full details of the instrumentation being carried aloft, and announcing an extension of the program to the UK and "other parts of the world."

Some of this was true—a weather research package *had* been developed for the U-2, and was to be carried on a special equipment bay hatch during training flights. NACA had a full-time U-2 project officer assigned, but the research agency had no control over where the flights were routed and received all of its data second-hand.[5] Watertown was now the "official" name for the Ranch. The CIA had named the secret base after the town in upstate New York which was the birthplace of DCI Allen Dulles.

For cover purposes, the first U-2 detachment which deployed to the UK was designated Weather Reconnaissance Squadron (Provisional)-1 (WRSP-1). Within the USAF, a provisional squadron did not have to file routine reports to higher headquarters.[6] The real designation was Detachment A, or Det A for short. When two further groups of operational U-2s were formed by the CIA in 1956, they were designated Det B (WRSP-2) and Det C (WRSP-3).

Det A was an unorthodox mix of USAF and CIA people, and civilian contract employees. General LeMay had ensured that his own men from SAC took most of the USAF slots, including the unit commander, operations officer, and their deputies. About a dozen more USAF officers and a similar number of enlisted men worked in operations, mission planning, provided support aircraft, and so on. There were about 40 CIA staffers looking after administration, security,[7] and communications. Since all of the maintenance—planes, sensors, life support systems—was done under contract, there were upwards of 60 more civilians assigned. Then there were the six "sheep-dipped" operational pilots. They weren't really sure who their boss was—the USAF Colonel who was officially the unit commander, or the GS-12 grade staffer from CIA who was the unit executive officer!

Back in Washington, there was a similar division of responsibility between USAF and CIA at Project HQ. The cover designation for the military personnel assigned there was the 1007th Air Intelligence Service Group, a unit of Headquarters Command (HEDCOM) in the Pentagon. Just as the U-2 deployment began, the military man who had done most to get the show on the road was reassigned, to Richard Bissell's great disappointment. Colonel Ozzie Ritland was promoted, and re-assigned to the Western Development Division, where he would become Bennie Schriever's deputy in the missile and satellite programs. His replacement as Bissell's military deputy was another USAF Colonel with an aeronautical engineering background, Jack Gibbs.

Reluctant Hosts

Det A's stay in Britain was short-lived. Shortly before it arrived, Soviet leaders Bulganin and Khrushchev paid a good will visit to the UK, arriving at Portsmouth dockyard on a Soviet Navy cruiser. Without permission, the British Secret Intelligence Service (MI6) dispatched a frogman to spy on the ship from beneath the water line. He was never seen alive again. The Soviets publicized the incident, an embarrassed Sir Anthony Eden dismissed Sir John Sinclair, the head of MI6, and told Parliament that the operation had been conducted "without the authority or knowledge of ministers." The political fall-out extended to Project AQUATONE. On 18 May, just days before the first overflights were due, Eden called Eisenhower and withdrew his permission for them.[8]

A disappointed Bissell told embarrassed British military and intelligence officials that a base in Germany was his second option. The Brits said they would allow Det A to remain at Lakenheath, but for training flights and maintenance purposes only. No overflights! Bissell and Johnson flew home, pondering their options.[9] Selwyn Lloyd told Eden that "should the active operations in Germany ever become the subject of any enquiry, the British government could deny all knowledge of them, since the presence of the special aircraft remaining at Lakenheath could be explained by reference to their meteorological mission."[10] Although Bissell and company may not have realized it at the time, this episode set the pattern for many future political constraints on U-2 operations. Even when they were persuaded to grant basing rights, foreign governments were reluctant to give unrestricted permission for illegal overflights.

On 11 June, Det A was moved to the busy USAF and CIA base at Wiesbaden, near Frankfurt. At Eisenhower's insistence, Bissell flew to Bonn in late June with CIA Deputy Director Pierre Cabell to get German permission for Soviet overflights. They were relieved to get an enthusiastic response from the "Iron Chancellor," Konrad Adenauer. Bissell decided to make a permanent home for Det A in Germany. Wiesbaden would be temporary, pending construction of suitable facilities at Giebelstadt airfield, 70 miles further east.[11]

The delay in starting overflights did mean that one major improvement could be made: the new and more reliable J-57-31 engines were shipped to Germany shortly after the planes arrived there. By mid-June Det A was ready for business. But President Eisenhower had not yet approved Soviet overflights, pending German approval and because an official U.S. delegation led by General Nathan Twining, USAF Chief of Staff, had taken up a Soviet invitation to visit Moscow. It didn't seem right to invade Soviet airspace while they were there!

The weather was good, so Bissell ordered a flight over the Soviet satellite countries instead. The detailed mission plan was transmitted from Project HQ to Wiesbaden, where Det A's operations staff drew up the flight charts and fuel curve graph that the designated pilot, Carl Overstreet, would take with him. A strict routine had been established for mission flights. The pilot would be sent to bed some 12 hours before the scheduled launch time, and woken with at least two hours to go. By then, Project HQ would have acquired the latest weather report and forecast for the target areas. If it was good enough, a go-signal would be transmitted by the Agency's own secure communications lines to the Det. The pilot would eat a high-protein, but "low-residue" meal. (The lavatory arrangements of the partial pressure suit were rudimentary, consisting of a bottle for liquid wastes which the pilot had to connect himself inflight, after much tugging at zips and hoses. However, some pilots found little call for the bottle even on long flights: the skin-tight pressure suit caused constant perspiration, and much body moisture was eliminated this way rather than through the kidneys. There was no provision for the disposal of solid waste from the body—a messy, uncomfortable, and embarrassing business if the pilot couldn't restrain himself).

After eating, the pilot would put on long white underwear, seams to the outside so that the tight pressure suit wouldn't impress them indelibly onto the occupant's skin. Now two physiological support technicians helped him into the suit itself for the two-hour prebreathing session. The suit was so restrictive of movement that a fellow-pilot would be detailed to perform most of the preflight checks at the waiting aircraft. He would also help the mission pilot strap himself in and connect all the hoses, wires, and so on. This "mobile" pilot would remain on duty throughout the flight. When the U-2 returned, he would "chase" it down the runway calling the height-to-touchdown over the radio, since the pilot's vision was badly restricted in the aircraft's landing attitude.[12]

Overflight

In the early-morning light on 20 June 1956, Overstreet climbed in his plane and taxied to the runway. Since complete radio silence was the rule, he awaited a green "go" light from the tower. Right on schedule, he took off from Wiesbaden on the first-ever operational U-2 mission. His biggest fear was the fear of screwing-up. He flew north and west to gain altitude, before looping back over the base and heading east. Mission 2003 penetrated denied territory where the borders of the two Germanys and Czechoslovakia met. This was by design. If communist radars in either country were detecting the flight, the neighboring countries might fail to co-ordinate, and lose contact. Overstreet had no indication that he had been detected, however, and Mission 2003 flew on across northern Czechoslovakia before turning north to pass east of Dresden and into Poland. The U-2 completed a tour of every major Polish city before exiting the way it came, flying over Prague and back into West Germany. No screw-up!

After landing, the undeveloped film and the SIGINT tapes were unloaded. After a quick check for quality control, they were transferred to a waiting plane, which immediately flew them back to the U.S. The CIA had made arrangements for Eastman Kodak to do the film processing. The original negative was developed and copied. Two duplicate positives were produced—it had been decided that these would allow better photo interpretation of high quality imagery, than working from paper prints.

It was two days before the film reached Washington. Former U.S. Navy photo-interpreter Art Lundahl had established a Photo Intelligence Division (PID) within the CIA in 1953. When Project AQUATONE was launched, Richard Bissell asked Lundahl to expand his small operation, which was housed in the temporary "M" buildings. Lundahl and his executive officer Sid Stallings moved the unit into the top four floors of a nondescript office block at 5th and K Street NW. It was owned by the Steuart Motor Car Company, which operated a car repair workshop on the ground floor. PID's new home was an unlikely location, in one of Washington's least-desirable suburbs. Soon, Lundahl's new recruits began to arrive, as did the new photo-exploitation equipment...microstereoscopes, X-ray light tables, and improved mensuration devices. Art Lundahl selected the codename AUTOMAT for his expanded operation.[13]

Lundahl's team examined the film from Mission 2003 on 22 June 1956, and compared it with film exposed over the U.S. during the U-2's operational readiness test the previous April. Although all the imagery had been taken by the interim A-2 camera rig, it was excellent quality. Much better than the old wartime German photography that was still the main reference to terrain behind the Iron Curtain! The A-2 system of three 24-inch focal length lenses in a trimetrogen configuration provided stereo overlapping imagery across a 36.5 nautical-mile swath of the ground.

The B-camera was more versatile, as well as higher resolution. If all seven of the "stop and shoot" lens positions were used, it could "see" from horizon to horizon. Since the position of so many U-2 targets behind the Iron Curtain was imprecisely known, this was a big bonus. But there had been development problems with the "B." On test flights, the moving lens assembly would bounce against the stop and cause the next image to blur. The problems weren't solved until late 1956. All the early U-2 overflights used the A-2 system instead.

How many Bombers?

After a week as guests of the Soviet Union, General Twining and his group left Moscow. At the Tushino Air Show, they had been treated to another spectacular fighter flypast, similar to those flown over the city in the last few Mayday parades. Over 100 of the still-new *Farmer* and *Flashlight* fighters, and at the very end—a flypast of even newer fighter prototypes, including a new version of *Flashlight*, three Sukhoi deltas, and two MiG deltas—forerunners of the MiG-21. Also flying by was the large, long-range four-engined turboprop bomber which had first been seen a year earlier, and nicknamed *Bear* by NATO. But there were only four of them, plus three of the big four-jet bombers codenamed *Bison*. The first of these had been spotted at the Fili airframe plant in Moscow's suburbs three years earlier, and a single example had flown over the MayDay parade in 1954. The following MayDay, ten *Bisons* had flown over in formation, suggesting that the bomber was already in service (the formation had circled to return over the parade a couple of times, leading some Western observers to double or treble-count them!). So how many *Bisons* and *Bears* were really operational? Twining's hosts were not forthcoming.

Neither did the Soviets talk about their missiles. At a conference in early 1956, British and American intelligence experts had pooled their knowledge of Soviet guided weapons development. It didn't amount to much. The only hard evidence came from German scientists who had been co-opted by the USSR in 1945 and obliged to work in Soviet weapons institutes. The USSR let most of them go in the early fifties, and they were thoroughly debriefed by Western intelligence. But the Anglo-American conference noted that "the German sources are now practically exhausted, and other means of obtaining intelligence should therefore have the highest priority."[14]

Another British intelligence estimate in mid-1956 noted that "large-scale firings of ballistic-type weapons are currently in progress" but admitted that "details of the rockets e.g. warheads are not known."[15] Those firings were taking place at the Kapustin Yar test range southeast of Stalingrad. Attempts to overfly this vital intelligence target had been made in the previous few years, with mixed results.[16]

By now, therefore, the intelligence community had a huge list of aerial reconnaissance targets behind the Iron Curtain. Bissell had established a new system for prioritizing them the previous December, by organizing the Ad-hoc Requirements Committee (ARC). CIA staff officer Jim Reber chaired this committee, where CIA, Army, Navy, and USAF representatives met to draw up the collection requirements. The top-priority targets were then passed to Project HQ, where mission planners devised potential routes for U-2 flights.

These mission planners knew that they had to cram as many targets as possible into each flight. President Eisenhower had read a detailed brief on Project AQUATONE, and told his staff secretary Colonel Andrew Goodpaster that he wanted all the vital targets covered as quickly as possible. On 21 June Bissell took James Killian and Din Land with him to the White House, where Goodpaster explained the President's thinking. Eisenhower was going to keep tight control of the flights, and wanted frequent reports. Despite all the CIA's assurances, Ike was worried that the flights might be detected, and that another balloon-type fiasco might ensue. SIGINT reports indicated to project HQ that the first overflight of Czechoslovakia and Poland had indeed been periodically detected by communist radars. But radar operators apparently did not obtain enough data to conclude that an unauthorized intruder was overhead.[17]

On 2 July, two U-2s left Wiesbaden in quick succession to fly over eastern Europe again. Jake Kratt flew Mission 2009 south across Austria and into Hungary. After Budapest, he turned south to fly along the Yugoslav border, then all the way across Bulgaria to the Black Sea. Glen Dunaway headed north on Mission 2010 before flying over East Germany, southern Poland, eastern Czechoslovakia and Hungary, and into Romania. This flight also reached the Black Sea before turning back across southern Romania, Hungary, and Austria. Each aircraft was airborne for nearly seven hours, and they covered nearly 6,000 miles of "denied territory" between them.[18]

Prime Targets

On 3 July, before the postflight analysis of those missions was complete, the President finally gave the go-ahead for U-2 flights over the Soviet Union. In Project HQ, Bissell chaired a final review of the mission plans. Everything looked good. Just before midnight, the go-signal was transmitted across secure transatlantic lines to Det A in Germany. Since Wiesbaden was six hours ahead of Washington time, it was 0600 on U.S. Independence Day, 4 July, when Det A pilot Hervey Stockman took off on Mission 2013 in the aircraft marked as NACA 187. He flew across East Germany and Poland and crossed the Soviet border near Grodno in Belorus. After flying over known bomber bases around Minsk, Stockman headed north for the prime targets on this flight, which were around Leningrad. But now he had company! Peering down through the driftsight, Stockman saw MiG fighters rising to try and intercept. Contrary to all American hopes, the flight had been detected! Stockman flew steadily on, putting his faith in Kelly Johnson's plane and the intelligence advice that said it couldn't be reached at altitude by the Soviet fighters. Sure enough, the fighters disappeared. The U-2 pilot flew over the naval shipyards at Leningrad where Soviet submarines were built, and the three known Soviet long-range bases around the city. Then Stockman turned west, to pass over more suspected Soviet jet bomber bases in the Baltic States before eventually landing back at Wiesbaden after an eight hour 45 minute flight.[19]

The next flight was planned to go all the way to Moscow. When asked to justify such a daring venture by Allen Dulles, Bissell told him that the first time would probably be the safest.[20] The pilot whose turn it was to fly next was told to get some sleep. With Independence Day parties going on all over the base, Carmine Vito didn't get much rest. At 0500 next morning, flying the same aircraft as Stockman had flown the previous day (Article 347), Vito took off on U-2 Mission 2014.

His route took him further south than Stockman had flown, over Kracow in Poland, then into the Ukrainian SSR to fly over Brest and Baranovici, where another of the suspected Bison bomber bases was located. There was considerable cloud cover, but it cleared away as Vito turned towards Moscow, virtually following the railroad from Minsk to the Soviet capital. As he crossed the Soviet border, Vito was aware that the air defense system knew he was somewhere above. The MiGs were on his trail, but they were mainly MiG-17s with a maximum altitude of 50,000 feet. Even the newer MiG nicknamed *Farmer* by NATO, which had just entered service, could only reach 55,000 feet.

Vito flew on over Orsa, where three *Bisons* had been reported earlier in the year by a ground observer. The rolling farmland seemed to stretch forever, divided into a mosaic pattern by the stone fences. He neared Moscow, where a new potential danger awaited. The USSR's first surface-to-air missile (SAM) system had been deployed around the capital. That much was already known to Western Intelligence, since U.S. and UK air attaches in Moscow had been monitoring its extensive construction all around the Soviet capital since late 1954. In March 1956, missiles had been seen at some of the sites, which numbered more than 20 in two concentric rings 20 and 40 miles distant from the city center.

The development of a SAM system around Moscow had aroused intense interest at the CIA's Office of Scientific Intelligence (OSI). Repatriated German rocket scientists had told OSI that the Soviets had been pursuing their wartime "Wasserfall" project, to be capable of knocking down aircraft as high as 65,000 feet. But the characteristics of the Moscow system suggested an entirely new line of development. The attaches had snatched glimpses of a strange double-disk rotating array for the missile guidance, which had earned the nickname Yo-Yo. This was nothing like a conventional parabolic antenna. Later, the "discs" were found to be triangular structures rotating at high speed. By the time that Vito's U-2 approached Moscow, OSI had commissioned a mock-up of Yo-Yo from a U.S. laboratory, in an attempt to determine its capability.[21]

Ahead of him, Vito could see the herring-bone pattern of the road layout around one of these SAM sites. He flew right over it and on over Moscow, where one of his main objectives was the Fili airframe plant where the *Bison* bomber was being built. The U-2 then rolled out to the northwest and headed for the science and technology complex just north of Moscow at Kaliningrad where Soviet missile development took place,[22] and the main Soviet flight test and research airfield at Ramenskoye. In so doing, Vito flew over two more of the SAM sites, but no missiles were fired.

Vito headed for home via his remaining targets: the Khimki rocket engine development and production factory; and more potential bomber bases along the Baltic. The cloud which had obscured some of the targets around Moscow thickened as he flew back, and it was overcast as he passed over the satellite countries. The weather was poor at Wiesbaden, and he needed a ground-controlled approach to land. Underlining the U-2's now-obvious detectability, he was picked up while still way over East Germany and at altitude by the radars which Canada operated as part of its 1st Air Division contribution to NATO.[23]

Detection

But Bissell's hunch about the Soviet air defenses around Moscow had proved correct: they weren't ready for him. In fact, an A-100 radar at Smolensk provided early warning for the Moscow SAM system, which the Soviets named the S-25 Berkut. This radar had detected Vito as he flew overhead on his way towards the capital, and the Soviet operators calculated that he was flying at 20,000 kilometers (65,000 feet). The radar alert proved futile, however, since the R-113 missiles for the Berkut system were not routinely kept at the firing sites. That same evening, Soviet Chief of the General Staff Marshall Sokolovskiy arrived at the Berkut command post. He was accompanied by a team of experts from the headquarters of the Voyska Protivovozdushnoy Oborona Strany (PVO—or Troops of the Air Defense Forces). They studied the target tracks of Vito's flight—and concluded that a radar malfunction or operator error had occurred. No aircraft could fly that high! The Berkut commander disagreed, and persuaded Sokolovskiy that the missiles should be armed, moved to the launch site, fueled, and placed on alert. It was done that very night.[24]

President Eisenhower had already made it clear he wanted the flights stopped if they could be tracked.[25] Exactly what constituted effective tracking was a matter of interpretation, however. It took some time for the CIA to correlate the SIGINT data from the U-2's own receivers with that obtained by the U.S.-operated SIGINT ground stations in Germany and the UK. Ike was told that the flights were being detected, though still imperfectly. He considered halting the operation, but told Goodpaster it could continue until more definite information was available.[26]

(Of necessity, the Soviet air defense system used High Frequency radio for rapid communication between its far-flung radars and command posts throughout the fifties. Since these signals propagated for long distances, the PVO communications network was easy prey to illicit eavesdropping from afar. When the U-2 flights started, James Killian happened to be inspecting one of the super-secret listening posts in West Germany operated by the National Security Agency (NSA). He watched as the American eavesdroppers tuned in on the Soviet panic. They couldn't figure out what was causing the flap, either. Despite their high-level SIGINT clearances, Killian didn't enlighten them on the U-2 project.)[27]

Meanwhile, bad weather prevented further overflights on the three days following Vito's flight. But on 9 July two U-2s took off in quick succession from Wiesbaden to resume the program. Marty Knutson flew Mission 2020 to the north of Berlin and up the Baltic Coast as far as Riga, then turned east and south to cover targets around Kaunas, Vilnius, and Minsk before returning via Warsaw. Carl Overstreet took Mission 2021 along the southern route via Czechoslovakia and Hungary, then turned northeast into the Ukrainian SSR. Flying as far east as Kiev, he made a high-altitude tour of more potential bomber bases...Gomel, Bobrusk, and Baranovichi again...before returning via southern Poland. Unfortunately, a broken shutter ruined much of the photography from one of these flights.[28]

Protest

The next day, 10 July, Glen Dunaway made Det A's furthest excursion yet to the East. He flew Mission 2024 all the way to Kerch on the eastern tip of the Crimean Peninsula, returning via Sevastopol, Simferopol, Odessa, and the satellite countries. Although Dunaway saw fighters beneath him over Odessa, the rest of the flight was uneventful. But while he was airborne, the USSR delivered a protest note about the previous days' flights to the U.S. Embassy in Moscow. After some debate over whether to concede that the flights had taken place, the Soviet leadership decided to swallow its pride.

The protest note described in some detail the "gross violations" of Soviet airspace "for purposes of reconnaissance," and ascribed them to "a twin-engined medium bomber of the USAF." It was clear that the Soviet air defense system had not only detected each flight, but had also tracked them for considerable distances. The PVO radars had not followed Stockman's mission all the way to Leningrad, however, and had identified three instead of two flights on 9 July. Vito's mission to Moscow had been mistakenly identified as two separate flights. Although the note admitted that one of these "had penetrated to a significant depth over Soviet territory," it did not mention Moscow. (Thirty-seven years later, two senior former Soviet air defense officers could still not bring themselves to admit that the Soviet capital had been overflown. Their published article on the U-2 flights over the USSR vaguely noted that the aircraft "got 1,000 kilometers deep into the USSR")[29]

U.S. intelligence concluded that the Soviets were indeed confused, and embarrassed to admit the true extent of the overflights. Handoffs of targets between air defense sectors were (as the U.S. had hoped) inadequate. It seemed that each sector was unwilling to be blamed for failure to intercept, and was therefore reluctant to transmit details of the intruder to the next sector. By referring to a "twin-engined bomber," the Soviets evidently couldn't believe that a single-engined plane could fly so high.[30] The Czechoslovak and Polish governments also protested the flights, though less specifically.

As far as President Eisenhower was concerned, the game was up. When the Soviet protest note reached the White House later on 10 July, U-2 overflights were suspended. Ike was annoyed that the assurances given by Bissell and others had proved false. They had told him that the flights would hardly be detected, let alone tracked.

But that was the wisdom of the time, within Western intelligence. A U.S. National Intelligence Estimate (NIE) issued while the U-2 was being developed noted that "although the Soviets have made great strides in radar development," their standard S-band V-beam early-warning radar, nicknamed *Token*, had no capability above 60,000 feet. The *Token* coverage was supplemented by two metric (VHF-band) air defense radars nicknamed *Dumbo* and *Knife Rest*, with similar capabilities. British intelligence thought that *Token* could detect bombers at 40,000 feet from 190 miles, but was less impressed with the two metric radars. They were right; the P-3 *Dumbo* and P-8 *Knife Rest* radars could only provide bearing and range. But the West knew nothing about the A-100 radar.[31]

During long-range training flights from the Ranch, U-2s had flown far to the north and within range of the DEW-line early warning radars, which managed to detect them periodically. A controlled series of flights were then made against Nike Hercules radar sites in southern California. The operators there couldn't assign accurate altitudes, tracks, or speeds to the high-flying bird, however. The CIA concluded that the PVO could not do any better, even in the western USSR, where Soviet radar coverage was most complete.

The CIA was half-right, but to Eisenhower detection was almost as bad as interception. He knew how seriously the Soviets would react to illegal overflights. After the Soviet protest note of 10 July 1956, Eisenhower would never again give carte-blanche for a series of U-2 overflights. He examined mission requests on a case-by-case basis, and wanted to know all the details regarding routes, defenses, and so on. More often than not, he withheld permission.

At Project HQ, Bissell quickly responded to the setback. He confirmed plans to deploy the second U-2 detachment in Turkey, close to the USSR's southern border where the PVO's radar coverage was thought to be weaker. He also turned to his trusted scientific advisors again, and asked them to determine whether radar-fooling devices could be added to the U-2 airframe.

Windfall

The "take" from the first U-2 flights was analyzed by Lundahl's team in the Photo Intelligence Division. Bissell and DCI Dulles also reviewed the imagery. They chuckled with amazement at the clarity of the black and white prints. Dulles rushed off to the White House with some samples. The President spread them out on the floor of the Oval Office and "we viewed the photos like two kids running a model train," Dulles later told Bissell.[32]

It soon became clear that the Iron Curtain had been thrust aside, in intelligence terms. To cope with the deluge of film and information, PID had to be expanded. More photo-interpreters from the Army and Navy were brought into the AUTOMAT project at the Steuart Building, though the USAF refused to assign its men there full-time. In particular, SAC jealously guarded its autonomy in photo-interpretation. The U-2 film contained many likely new Soviet targets for its bombers. SAC headquarters maintained the USAF's largest photo-interpretation shop, and insisted that a duplicate negative and a duplicate positive of the U-2 film was sent to the 544th Reconnaissance Technical Squadron (RTS) at Offutt AFB. A distinct rivalry developed between SAC and AUTOMAT over the ensuing years.[33]

Within the PID, one group of photo-interpreters was assigned the military targets, and another the industrial targets. A third group compiled an index—a vital role given the sheer volume of film. The initial "readout" of target information was compiled into an "OAK" report, for quick circulation to those who were cleared to receive intelligence derived from the U-2 flights. Strict new control procedures were introduced to protect the source, which was classified above Top Secret by means of additional codewords. The U-2 imagery itself was codenamed CHESS, and the intelligence derived from it was codenamed TALENT.

After the OAK report was done, the PIs settled down for a more considered, second-phase study of the film. At first, they used paper prints derived from the original negative for interpretation, but it was soon realized that the high-quality U-2 imagery could best be interpreted from the duplicate positives, which gave a better gray scale. Film from the U-2's 70mm tracker camera, which ran continuously throughout the flight, was also examined. It provided positional reference, especially when the aircraft had flown over featureless terrain, and was very valuable to the photogrammetrists within PID.[34]

The big news from the first U-2 overflights was that no massive Soviet bomber build-up was evident. There were plenty of *Badger* medium-range bombers on the film, mostly based in European Russia. They posed a threat to Western Europe, but not to the U.S., unlike the supposedly long-range *Bisons* and *Bears*. But none of these were seen deployed. The current NIE on Soviet attack capability suggested that about 40 *Bisons* and 35 *Bears* were operational now, and predicted 470 in service by mid-1958 and 800 by mid-1960. After economic analysis of the Soviet industrial base—to which the U-2 imagery also contributed—the CIA concluded by late 1956 that these projections were unreal. There was no "bomber gap."

Although the USAF disagreed, the CIA's analysis was correct. The Myasishchev design bureau had been struggling to make the M-4 Molot (*Bison*) into a true strategic bomber with intercontinental range. The ten aircraft which had been seen by Western air attaches in the 1955 MayDay flypast were interim versions, but even the 3M version (*Bison-B*) built in 1956 was far from satisfactory. As it transpired, there were never more than 60 M-4s put into service, and the type was only ever produced at one plant—Fili. As for the *Bear*, Tupolev's large but conservative Tu-95M design certainly had intercontinental range, but no great speed. Unlike the M-4, the Tu-95 would remain in production for many years, but at a relatively slow rate.

The estimates of Soviet long-range bomber strength were formally revised downwards in 1957-58, although the USAF continued to take the bomber gap seriously. The film product of the U-2 missions was so highly compartmented that even within the intelligence community, many were not aware of this highly reliable new source of information. For instance, drawings of the important targets revealed by the U-2 flights would be passed to the United States Air Forces in Europe (USAFE), rather than the imagery itself. USAFE target planners would superimpose the new data on their old German wartime imagery. Often there were conflicts, as the target planners protested that famous hunting lodges or hotels were located where the new data indicated runways. Even so, the U-2 imagery represented an intelligence windfall for USAFE, which soon began filling its target folders with the new or revised locations.[35]

Suez

In late August 1956, Detachment A was alerted for more operational missions. This time, though, they were to fly south instead of east. The targets were in the eastern Mediterranean, and were prompted by the Suez Crisis. The increasingly pro-Soviet President Nasser of Egypt had nationalized the Suez Canal in late July.

After a month of tense negotiations, British and French troops set sail for Cyprus on 29 August. On the same day, two U-2s departed Wiesbaden and flew eight-hour missions across the entire area before landing at Incirlik, the USAF base which was being prepared for Det B in Turkey. The next day, the two aircraft flew similar missions before landing back at Wiesbaden. The imagery showed the military preparations of various parties to the dispute, but it was only made available to one of them. On 7 September, Jim Reber from ARC and Art Lundahl from PID flew to London and briefed British intelligence officials. Unbeknown to President Eisenhower, who did not believe that his closest ally would take military action over Suez, the British subsequently used the U-2 imagery which Reber and Lundahl left behind to identify landing sites, drop zones, and invasion routes.[36]

The crisis was deepening, and the CIA decided to use the U-2 on a regular basis to monitor the situation. To speed up the interpretation of imagery, Lundahl made arrangements for the USAFE's 497th Reconnaissance Technical Squadron (RTS) at Wiesbaden to process the U-2 film, and he sent some staff from AUTOMAT do the initial readout from the negatives there. Fortunately, the CIA's second U-2 group was just now moving into Turkey, and would be close to the area of interest. Following approval for a U-2 base by the Turkish Prime Minister in early May, Leo Geary and the CIA station chief in Turkey surveyed various airfields before recommending Incirlik, near Adana in the isolated southeast of the country. Det B began moving from the U.S. in late August.

The CIA's second U-2 group had not been accident-free during the work-up period at the Ranch. On 15 May, Wilbur Rose became the first U-2 pilot to be killed when he stalled the aircraft while turning base leg. Rose had just taken off with a full fuel load, but a pogo had "hung up," and he was returning to try and shake it loose. But he let the bank angle increase too far, and the fuel-laden right wing just kept on dropping.[37] Two weeks later, another pilot ran out of fuel on final approach to the lakebed. The aircraft was damaged, but repairable.

The third group arrived at the Ranch for training in August, but one of its pilots (Frank Grace) was soon killed. During a night takeoff on 31 August, he became disoriented, lost control, and stalled in. Det B eventually deployed to Incirlik with eight pilots in early September. A ninth pilot who had trained with them, Howard Carey, was sent to Det A. Shortly after he arrived, Carey became the third U-2 fatality. On 17 September he lost control sometime during the climbout, and crashed.[38]

The U-2 had been built, flight-tested, *and* deployed in an unprecedented 17 months. It was hardly surprising that some things weren't right yet. After the fatal accidents, some modifications were made. The solenoid release valve for the pogos was replaced by a system which allowed the pogos to drop away automatically upon takeoff; the elevator tab was extended to provide more sensitive trimming; and a gyro-compass was added as another aid to navigation.

Det B flew its first operational mission over the eastern Mediterranean on 11 September, and its second on 25 September. Det A also flew nine missions to monitor the Suez Crisis this month, as U.S. Secretary of State John Foster Dulles (the brother of CIA's Allen Dulles) became increasingly suspicious of British, French, and Israeli intentions. One U-2 flew over the French naval base at Toulon during its delivery flight to Det B from Germany to Turkey. The British islands of Cyprus and Malta in the Mediterranean were regularly re-photographed; the CIA's photo-interpreters counted the number of tents that sprang up as British troops were deployed there. They also noted the Hunter fighters and Canberra bombers arriving from the UK as the build-up continued. Det B flew another nine missions in October, and some of these revealed nearly 60 Mystere IV fighter-bombers on Israeli airfields, whereas France had admitted supplying only 24.[39]

Foster Dulles described the Suez invasion which began on 29 October as "a collusion and deception" against the U.S. by the UK, France, and Israel. But thanks to the U-2 photography, plus SIGINT intercepts and the diplomatic comings and goings in Europe, the U.S. was not caught unawares. Once the fighting started, U-2 pilots had a grandstand view from 70,000 feet through their driftsights, since flights were mounted from Incirlik every day. The most notable of these was flown by Bill Hall on 1 November, when he passed twice over Almaza airfield near Cairo. During the first pass, when he was heading west, rows of Egyptian fighters were captured on film. The second pass was only a short time later, after Hall reversed course to head east again. This time, though, the Egyptian fighters were smoking wrecks. British and French bombers had attacked the base between U-2 passes. President Eisenhower was shown the before-and-after photographs, and is said to have remarked "Twenty-minute reconnaissance—now that's a goal to shoot for!"[40]

Det B continued to fly over the region, although less intensively after the British and French agreed to a cease-fire on 7 November. To shorten the intelligence cycle still further, PID set up a film-processing unit at Incirlik itself. In the subsequent three years, Middle East missions became a "milk run" for Det B, since they usually went undetected and unopposed.

Detection - again
But the main intention of basing U-2s in Turkey was to overfly the USSR. U.S. intelligence estimated that the PVO had limited radar coverage along its Central Asian border, and fighter bases were few and far between. If the U-2 could penetrate from this direction undetected, it might not subsequently be "tagged" hostile, even when it approached more heavily-defended areas. President Eisenhower was well aware of this thesis, but despite the combined pleadings for more flights by Bissell and the chairman of the Joint Chiefs of Staff, Admiral Radford, he remained cautious. Eventually, he approved a few overflights of the satellite countries and a flight into the southern USSR, which "should stay as close to the border as possible."[41]

On the morning of 20 November 1956, Det B pilot Frank Powers took off from Incirlik and flew into Syria and Iraq. He passed over Bagdhad and into Iran before turning north towards the Caspian Sea. He crossed the Soviet border and flew over Baku before turning west to overfly Yerevan. The flight was then supposed to head for Tbilisi, but the aircraft's electrical inverter malfunctioned, forcing Powers to make an early return to Incirlik. Mission 4016 was one of the first to use the B-camera, which returned some imagery of the border airfields. But the flight had not gone undetected. Soviet fighters had once again been scrambled; Powers had seen their contrails.[42]

A month later, Det B flew its first dedicated SIGINT mission along this same border, staying outside Soviet airspace, but extending all along the Soviet coastline from the Black Sea to the Caspian Sea and beyond. This was an attempt to determine the Soviet radar "order-of-battle" for the region in greater detail, using the newly-developed System 4 sensor package. Designed by Ramo-Wooldridge, it weighed over 500 lbs and completely replaced the camera sensor in the Q-bay. Even so, it was much smaller than the bulky SIGINT equipment previously flown along Soviet borders by converted U.S. bomber and patrol planes. Those systems required a crew of airborne "ferrets" to tune the receivers and direction-finding (DF) equipment, and record the intercepted signals. System 4 couldn't do DF, but it was a significant advance in automation, so that all the read-out and analysis could be done on the ground after the flight returned. U-2s continued to carry the smaller and less capable SIGINT Systems 1 and 3 in the nose, when photography was the primary mission. System 1 provided electronic intelligence (ELINT), but its crystal video receivers had to be preset to certain radar bands before takeoff, and only Pulse-Repetition Frequency (PRF) and scan rate were recorded. System 3 was a COMINT system using three swept-frequency receivers to record VHF transmissions, such as those between Soviet fighters and their ground controllers.

Yet another sensor system was available for the U-2. Detachments A and B each had a single aircraft equipped with the Westinghouse APQ-56 side-looking radar (SLR). This radar had been commissioned by SAC in 1954 as an extra means of identifying and locating airfields behind the Iron Curtain. It used a wide band receiver and a special recorder, and produced mapping images which were higher resolution than the "standard" Plan Position Indicator (PPI) airborne radars. It was fitted to ten of SAC's RB-47Es. These also carried cameras, and made a number of reconnaissance penetrations into Soviet border areas in 1955-56. Although the antennas were long, they were not heavy, and a time-shared version was soon developed which used only one transmitter and receiver, switching them back and forth between the left and right-looking antennas.

This further reduced the weight, and since this SLR operated in the rarely-used Ka-band using a narrow antenna beamwidth, the danger of its signal being intercepted by Soviet defenses was slight. According to SAC, these factors made it a practical proposition for the U-2. The antenna was installed in a faired radome below the U-2's rear fuselage, and a cathode ray tube (CRT) plus film recording device was fitted in the cockpit. The pilot had to make occasional adjustments to the radar image, and this proved awkward since the CRT was located behind his right elbow! He had to use a mirror to see the image. But compared with the U-2's superb photo imagery, the SLR had little to offer. The resolution was no better than 40 feet. The aircraft had to be flown in a straight line to ensure the correct map geometry, and a RADAN navigation system had to be added to help the pilot maintain a precise course.[43]

On 10 December 1956, Dets A and B flew over Albania, Bulgaria, and Yugoslavia. President Eisenhower had authorized these flights a month earlier. Next day, though, the USAF flew three RB-57Ds straight over Vladivostok at lunchtime on missions which the President claimed he knew nothing about.[44] Six days later, the USSR delivered a strong protest which noted that the violation had occurred in such clear weather as to "exclude the possibility that the pilots were lost." A furious Eisenhower told JCS Chairman Admiral Radford and DCI Dulles to stop all further U.S. reconnaissance flights over Iron Curtain countries.

The RB-57D was one of the aircraft sponsored by the Bald Eagle project. Despite its known deficiencies, compared with the U-2, the USAF had proceeded with the rewinged jet bomber. It entered service with the 4080th Strategic Wing at Turner AFB, GA, which (like the U-2) inherited personnel from SAC's disbanded fighter wings. The flights over Vladivostok were part of the first operational deployment of the "D-model," which could reach 65,000 feet—on a good day. The aircraft were temporarily resident at Yokota, the USAF airbase near Tokyo.

The CIA was planning to base its third U-2 group in Japan. Det C began training at the Ranch in mid-August 1956, and was ready to deploy by year-end, despite losing one aircraft in an accident on 19 December. The deployment was held up by diplomatic problems with the Japanese. Det C eventually deployed to the U.S. Naval Air Station at Atsugi, west of Tokyo, in mid-March 1957.[45]

The 4080th wing was also scheduled to receive the 29 U-2s which had been ordered for $32 million by the USAF early in 1956.[46] In August, the wing's first cadre of maintenance troops was sent for training to the Lockheed factory at Oildale, CA, 90 miles north of Burbank. The USAF U-2s were being built here, rather than in Building 82, which had produced the 20 aircraft for the CIA. Lockheed had previously used the Oildale factory for sub-assembly work. Art Westlund ran this outpost, recruiting farm workers as well as out-of-work Korean War veterans with some airplane experience. The finished articles were disassembled and wheeled onto the adjacent Bakersfield Airport, from where C-124 transports airlifted them to the Ranch. The first USAF aircraft, Article 361, arrived at the still-secret base in September 1956. The first USAF pilot, Colonel Jack Nole of the 4080th, was checked out on 13 November 1956. USAF headquarters assigned the unclassified nickname DRAGON LADY to the project.

As the USAF's DRAGON LADY project got underway, General LeMay lobbied again for control of the whole U-2 program. Thanks to Eisenhower's ban on overflights of Iron Curtain countries, there wasn't much activity in the CIA detachments, except for a few flights over Albania and Yugoslavia which Project HQ had authorized (presumably because it defined those two countries as being outside the Iron Curtain). Bissell wrote a memo to Dulles, justifying the retention of the Agency's U-2 operation...greater security, deeper cover, civilian pilots, and so on. The issue was thrashed out at a White House meeting on 6 May 1957, where the President rejected the USAF takeover bid.[47] LeMay had to be content with a secondary role for "his" U-2s—nuclear fallout sampling in the upper atmosphere.

Eisenhower's decision to keep the CIA in the U-2 business was partially based on the prospect of renewed overflights, which the President's own Board of Consultants on Foreign Intelligence Activities had recently recommended. The Soviet threat was switching from bombers to guided nuclear weapons, and there were huge gaps in Western knowledge. But there was another reason why Eisenhower was prepared to consider more flights. Dick Bissell said he had a new way to prevent the flights being detected.

Notes:

[1] DEFE7/2102 and DEFE10/474 files in UK Public Records Office (hereafter PRO files)

[2] PRO file AIR19/826

[3] as above

[4] Johnson diary, 17 May 1956

[5] Knowledge of the upper atmosphere was indeed still in its infancy, as the early U-2 pilots sometimes found to their cost during flights from The Ranch. One was flying at 68,500 feet near Reno when the aircraft rapidly climbed 2,000 feet, almost exceeding the g-limits. It had been pushed upwards by the Sierra Wave, which meteorologists at the time believed was thoroughly dissipated above 45,000 feet. Another encountered unexpectedly strong head-on jetstreams after a flameout, which destroyed his ability to glide to the nearest available runway for a deadstick landing. Fortunately, he eventually

managed to relight his engine - just above the rim of the Grand Canyon! However, NASA eventually concluded from the VGH recorder carried on the U-2 that clear-air turbulence was a rare occurrence above 60,000 feet. See NASA Technical Note D-548, October 1960

[6] Pedlow/Welzenbach, p94

[7] the security officer for Det A was Edmund P. Wilson, who would later achieve notoriety as an illicit arms dealer, and be convicted and sent to jail for conspiracy to murder

[8] Johnson diary, 18 May 1956

[9] Johnson diary, 19 May 1956

[10] PRO file AIR19/826

[11] Pedlow/Welzenbach, p95/97. This contradicts the Bissell article, where he recalls visiting the German Chancellor *before* the aircraft were moved to Germany

[12] As the pilots gained experience, the CIA detachments dispensed with the need for a 'mobile' chase upon landing. But the USAF deemed this procedure essential, and it continues to this day, no matter how experienced the mission pilot.

[13] Dino Brugioni, "Eyeball to Eyeball", Random House, NY 1990, Bissell book p104, and Bill Crimmins interview. According to Lundahl, interviewed for the Bissell book, the inspiration for the AUTOMAT codename came from the convenience stores in New York city, which were open around the clock, and where various items could be purchased from automatic vending machines. Because a classified codename could not begin with a vowel, the initials of Lundahl's security officer Henry Thomas were prefixed to the name, hence HTAUTOMAT.

[14] UK Joint Intelligence Committee (JIC) report (56) 23, in PRO file CAB158/24

[15] JIC report (56) 32 in CAB 158/24

[16] exact details are still not known or declassified. see footnote 15 to chapter one. According to "The Slick Chick" by Jim Tuttle, in Journal of the American Aviation Historical Society, Summer 1997, a specially-modified F-100A Supre Sabre fighter capable of flying over 50,000 feet, flew from Turkey to photograph Kapustin Yar in 1954. The date cannot be correct, since the three aircraft didn't reach Europe until May 1955. However, they flew 16 top-secret missions between then and the arrival of the U-2 in June 1956, according to the history of USAF's 7499th Support Group.

[17] Pedlow/Welzenbach, p101

[18] map of flights from Lockheed Skunk Works Star, 24 May 1996

[19] most flight data from Hervey Stockman, interviewed by author

[20] Bissell article

[21] Charles Ahern, "The Yo-Yo Story: An Electronic Analysis Case History", CIA Studies in Intelligence, Volume 5, Winter 1961, declassified 1997

[22] this is the Kaliningrad north of Moscow, later renamed, and not the port city on the Baltic which still bears this name

[23] some flight data from Carmine Vito, interviewed by author

[24] Col-Gen Yu.V.Votintsev, "Unknown Troops of the Vanished Superpower", Voyenno-Istoricheskiy Zhurnal, No 8, 1993, hereafter Votinsev article. The CIA's Office of Scientific Intelligence (OSI) eventually concluded in 1958 that the S-25 system (designated SA-1 and nicknamed GUILD by the West) was a major technological advance with good capability at low-medium altitude against multiple targets. see Ahern article. However, the USSR chose to develop mobile SAM systems instead, and the costly, fixed-site S-25 system was not deployed elsewhere.

[25] Pedlow/Welzenbach, p106

[26] as above, plus Goodpaster's Memorandum for Record, 10 July 1956, White House Office of the Staff Secretary, Alpha series, in the Eisenhower Presidential Library, hereafter Goodpaster memo plus date

[27] James Killian interviewed by Michael Beschloss, quoted in the latter's "Mayday: Eisenhower, Khrushchev and the U-2 Affair", Harper and Row, NY 1986

[28] Pedlow/Welzenbach, p108

[29] G.A.Mikhailov and A.S.Orlov, "Mysteries of the Closed Skies", New and Newest History, Moscow, June 1992, hereafter Mikhailov/Orlov. Welzenbach/Pedlow p109 mistakenly conclude that the USSR did not realize that Moscow had been overflown.

[30] the Soviet Union was probably confusing the U-2 with the USAF's RB-57A, which was now based

in Europe. This mistake allowed the US to formally deny that the USSR had been overflown. "No US military planes...at the time of the alleged overflights could possibly have strayed so far from their known flight plans," was the State Dept's formal response to the Soviet protest.

[31] NIE 11-5-55 dated 12 July 1955; British estimates from JIC papers in PRO files CAB158/23 and CAB158/24.

[32] Rich and Janos, p165

[33] Wayne Jackson, "Allen Welsh Dulles as DCI," a CIA history declassified in 1994, Volume II, p28-30

[34] Bill Crimmins interview

[35] Lt Gen Eugene Tighe, "Imagery and Reconnaissance Reminiscences" in American Intelligence Journal, Winter/Spring 1992

[36] Nigel West, "The Friends: Britian's Post-War Secret Intelligence Operations", Wiedenfelt & Nicholson, London 1988, p112

[37] crash details from Robertson interview

[38] Like all the accidents which befell CIA U-2s, the findings of the investigation have not been declassified. It is known that two Canadian F-86 interceptors were flying in the vicinity. Carey might have been trying to avoid them and lost control, it was surmised. According to Bissell, book p108, Kelly Johnson determined that the cause of this accident was overpressure in the wing tanks as the aircraft approached operating altitude after a very steep climb. This led to structural failure of the wing. The solution was to install a simple relief valve. However, members of the original U-2 flight test team insisted to this author that the overpressure problem was identified and solved a year earlier during the very first flight tests. The Lockheed U-2A flight test development report (SP-109) shows that to correct an unacceptably low rate of fuel transfer from the wings to the sump tank, the original system of unpressurized tanks relying on a gravity feed system was replaced by pressurization to 1.5psi by engine compression bleed air. However, a Service Bulletin dated November 1958 did make changes in the fuel system "to prevent excessive pressure in the wing tanks". An alternative explanation for Carey's crash was offered by fellow pilot Marty Knutson (interview). This is that there was an uncommanded wing flap extension, which led quickly to loss of control.

[39] Pedlow/Welzenbach, p114-6; Charles Cogan, "From the Politics of Lying to the Farce at Suez: What the US Knew" in Intelligence and National Security, Vol 13, No 2 (Summer 1998), Brugioni p33.

[40] Brugioni, p34. Another version of this story, from the Amory interview, JFK Library, attributes a similar remark to an RAF intelligence officer who was also shown this imagery. But Pedlow/Welzenbach note, p114, that the only U-2 imagery from the Suez Crisis that was shown to the UK, was during the 8 September briefing mentioned earlier.

[41] Pedlow/Welzenbach, p124

[42] Pedlow/Welzenbach, p124 states that a secret Soviet protest was made after this flight, but none has been revealed by State Dept archives

[43] radar details from Bill Griffiths, former Westinghouse engineer, in letter to author.

[44] During 1955-56, a substantial number of penetration overflights of the USSR and Eastern Europe were carried out by SAC RB-47Es, and by specially-modified RB-57As and F-100As assigned to USAFE and PACAF. In recent years, historians employed by the US government have claimed that these flights were always performed under an authority properly delegated by the White House, through the JCS, to the commanders of SAC, USAFE and PACAF. See for instance, "Cold War Overflights: Military Reconnaissance Missions over Russia Before the U-2" by Cargill Hall, in Colloquy, April 1997. This author has not yet seen any declassified memoranda to support this claim. Whatever the truth, it is certain that the 'Vladivostok incident' ended such military flights over the USSR. From now on, the White House approved only CIA U-2 flights, and these only sparingly.

[45] personnel from Det C told the author that the US government assured Japan, that no overflights would be conducted from Japanese bases by the U-2. According to Pedlow/Welzenbach, p134, Japan "had no control over US military bases in Japan," and the hold-up was caused by the difficult search for accommodation in the area.

[46] The USAF subsequently ordered a 30th aircraft - see Appendix Five

[47] Pedlow/Welzenbach, p127-8

4

Avoiding Detection

After the original series of U-2 overflights were detected, and therefore stopped by the White House, Dick Bissell consulted his scientific advisors again. Din Land's group repeated their earlier advice, that the U-2 would only be able to out-fly the Soviet air defenses for a couple of years. In August 1956, therefore, Bissell and his military deputy, Colonel Jack Gibbs, began searching for a successor to the Angel. The task would take them to research laboratories and aircraft companies all across the U.S. Gibbs, a Caltech graduate, was amazed at Bissell's ability to master aerodynamic theories without having an academic grounding in the subject.[1] From unlikely beginnings, the DPS was emerging as a new government "sponsor" of leading-edge aerospace technology.

Bissell was indeed a "Renaissance man." Before joining the CIA full-time in 1954, a stint at the Massachusetts Institute of Technology (MIT) brought him into contact with mathematicians, electrical and aeronautical engineers, and physicists. Bissell now tapped these contacts, especially within MIT's Radiation Laboratory (RadLab), and asked them if the U-2 could be modified to make it less vulnerable to radar detection.

In mid-August, Kelly Johnson traveled east for a meeting with Bissell and some of the Boston scientists. Ed Purcell, a physics professor at Harvard University and a wartime pioneer of microwave research at the RadLab, was briefed on the U-2's overflight problems, and related his "mirror" theories on radar deception. The RadLab had already designed two types of resonant radar absorber, though their main application had been to reduce electromagnetic interference, rather than radar cross-section. The small group brainstormed into the small hours, and reconvened at 7 am. By midday, they had devised a program to apply radar-canceling devices to the U-2. Project Rainbow—the first-ever attempt to make an operational aircraft "stealthy"—was underway.[2]

An Invisible U-2?

It was possible to apply a resonant absorber to the U-2 to fool the USSR's *Token* early warning radars, it was thought, since they operated at microwave frequencies around 3 GHz. The absorbers worked on the quarter-wavelength principle: a thin resistive sheet would be placed one-quarter wavelength from the aircraft's skin, and would reflect some of the incoming radar energy. The remaining energy would strike the U-2's metal skin and be reflected, but these "emergent waves" would cancel out the reflected waves, being 180-degrees out of phase with each other. To defeat the 10-centimeter wavelength of the *Token*, the absorber needed to be "only" some one-inch thick. But it would probably have to be applied to the whole of the lower half of the aircraft—wherever those radar waves might strike.[3]

But the metric-wavelength Soviet radars posed a greater problem! In addition to the *Token* (which the USSR designated the P-20 Periskop), the Soviet early-warning net still relied on numerous P-3, P-8, and P-10 static radars operating in the 65-86 MHz frequency range. These radars, nicknamed *Dumbo* and *Knife Rest* by the West, had also detected the U-2, though unlike the *Token* they had no inherent height-finding capability. Defeating their 3-4 meter wavelengths called for a different solution, and one that did not appeal one jot to Kelly Johnson. RadLab proposed that small-gauge wire with precisely-spaced ferrite beads be strung around the aircraft. If they, too, were placed at quarter-wavelength, they might trap and defeat the metric radar waves.

Bissell prevailed over Johnson, and both methods of making the U-2 "invisible" were given the go-ahead. The practical work was contracted by the Scientific Engineering Institute (SEI), a "front" company set up by Bissell so that the CIA could write formal contracts with the scientists in Boston. The key scientist who took Purcell's theories and applied them to the U-2 was Dr. Frank Rodgers of MIT. At the Skunk Works, the Rainbow project was led by Luther MacDonald, assisted by Mel George and Ed Lovick. They had to devise a mounting system for the wires, and adapt the laboratory-standard absorbers for the rigors of flight. Eventually, two absorbers would be used: the Salisbury Screen grid consisting of graphite on canvas, and Eccosorb, an elastomeric type which could be tuned to more than one frequency.

Death in the Desert

By mid-December, the first U-2 in a Rainbow configuration was being test-flown against radars at Indian Springs AFB, on the Nevada range. These were operated by Edgerton, Germeshausen & Grier (EG&G), another offshoot of the MIT RadLab. The wires were mounted on laminated wood poles attached all along the leading and trailing edges of the wings and horizontal tail. There were further attachment points at the nose, and even above the cockpit, meaning that the last wires had been secured *after* the pilot had climbed in. Not surprisingly, the test pilots hated them. They flapped in the slipstream, and were a potential hazard to flight control if one should break.

Throughout the first few months of 1957, the flights proceeded with much trial and error. The wires were cut and spaced to different wavelengths, and assembled in different configurations. The various absorbers were tried out, along with ferrite-based paints. The results were mixed, but some success was achieved against the Indian Springs radars.

But the weight and drag penalties were significant; they reduced the U-2's maximum altitude by at least 5,000 feet, and range by 20%.

On 2 April 1957, Lockheed test pilot Bob Sieker was flying the prototype U-2 on yet another Rainbow test when the engine quit at 65,000 feet. As his pressure suit inflated, the faceplate clasp on Sieker's helmet failed, and the pilot quickly lost consciousness. The aircraft went out of control and entered a descending flat spin. As it reached the lower atmosphere, the hypoxic Sieker recovered enough to realize he must get out of the plane! Struggling against the g-forces, he eventually managed to discard the canopy and climb out. But it was too late. Still in a flat spin, the aircraft hit the ground virtually intact, bounced, and a small fire broke out. It took rescuers nearly four days of searching from the air to find the crash site, in the remote desert about 40 miles north of the Ranch. They found Sieker's body less than 200 feet from the wreckage. His parachute pack was opened, but the chute was only partially deployed.[4]

A post-mortem on Sieker at Tonopah revealed his hypoxia, and since the aircraft was virtually intact, it wasn't too difficult to work out the root cause of the accident. The radar absorbent coatings had acted as insulation around the engine bay, causing the hydraulic system to overheat and reduce pressure to the fuel boost pump motor. Additional cooling was added to correct the problem. But if his life support system hadn't failed, Sieker could have still brought the plane home. Kelly Johnson called for a redesign of the faceplate, a dual oxygen regulator, and an ejection seat that could be used on an interchangeable basis with the existing seat. At least two U-2 pilots had strongly recommended a dual oxygen system more than a year earlier, and the accident in which Det C pilot Bob Ericson had lost control and been forced to abandon a U-2 the previous December, was also caused by hypoxia, induced by a faulty oxygen feed.[5]

Goodbye to the Ranch
At the time of Sieker's accident, training activity for the USAF's own U-2 operation had reached a peak, with 18 pilots plus maintenance crews at the Ranch. Unfortunately, operations at the secret base were being disrupted by renewed nuclear testing at the AEC range right next door. The PLUMBBOB series of atmospheric

Clarence 'Kelly' Johnson, Lockheed's brilliant aeronautical engineer, was 44 when he started design of a high-flying reconnaissance aircraft in early 1954. (Lockheed AG2149)

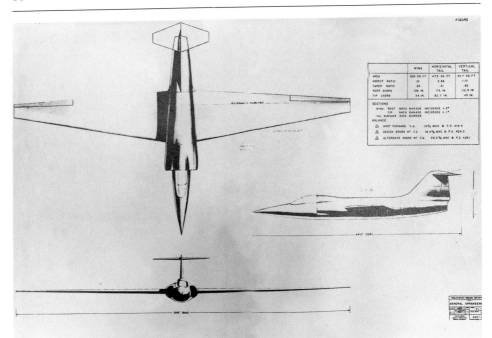

The CL-282 was Kelly Johnson's attempt to adapt the XF-104 for the spyplane requirement. A new, 70-foot wing was the key feature, along with radical weight-saving ideas. By early 1955, the U-2 had evolved from this radical design, but with many changes. (from Lockheed Report LR9732)

Edwin Land was a brilliant scientist whose drive and vision forged the Polaroid company and made him a natural choice as a top-level advisor to the White House. More than anyone else, he ensured that the U-2 was actually built. (Polaroid Corporate Archive)

Two key designers of the U-2's optical systems confer. Dr. Jim Baker (left) was the Harvard astronomer and expert lensman who conceived the B-camera, and served on the Land Panel which urged development of the U-2. Richard Perkin (right) was president of Perkin-Elmer, the company which produced the camera lenses, and also the U-2's tracker camera and driftsight. (via Cargill Hall)

The 115-foot wingspan of the Bell X-16 is readily apparent on this model, placed next to an F-86 Sabre for comparison. Funded by the USAF, development of the X-16 was well underway when it was canceled in favor of the CIA's U-2. (via Jay Miller)

Pratt & Whitney struggled to adapt the J57 turbojet for the very high altitudes, but the -31 version eventually proved to be effective and reliable. (via Jay Miller)

Richard Bissell stands close to an enduring symbol of the Cold War, Berlin's Brandenberg Gate. Bissell ran the U-2 project at the CIA, bringing his formidable intellect and a free-wheeling management style to bear. (via Frances Pudlo)

Colonel Ozzie Ritland was the Pentagon's first liaison officer to the U-2 project, and soon became Bissell's military deputy at the CIA itself. He returned to the USAF in May 1956, but was to become Bissell's deputy again in 1958, for the CORONA reconnaissance satellite. (AFFTC History Office)

Work proceeds on a U-2 wing in the assembly jig. The highly-polished leading edge has already been fixed to the front spar. (Lockheed PR-1039)

Colonel Leo Geary was in charge of USAF liaison to the CIA's air operations for an unprecedented ten years, 1956-1966. He became an intimate of Kelly Johnson, as the Lockheed Skunk Works provided first the U-2 and then the A-12 Blackbird under streamlined, super-secret conditions. (NARA)

Tony LeVier, Lockheed's chief test pilot for the F-104, was chosen by Kelly Johnson to fly the U-2 prototype. "I switched from flying the plane with the shortest wings in the world, to the one with the longest!" he said. (Lockheed AL8068)

The prototype "Angel" or "Article" stands on the hastily-laid tarmac at The Ranch. The strange new plane had been flying for three months, before it was formally designated as the U-2. (Lockheed 84-CC-3284)

The early U-2 cockpit with its basic flight instrumentation, and the large yoke and control wheel. The rubber sighting cone for the combined driftsight/sextant display is also prominent. (Lockheed PR1627)

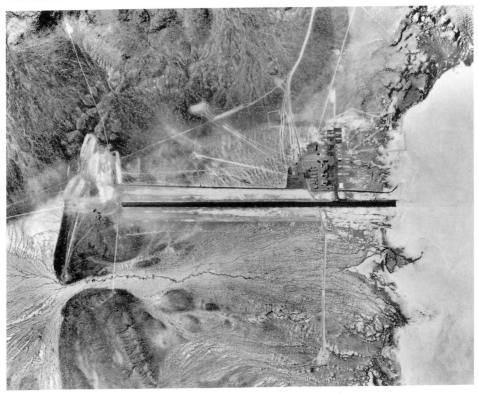

Overhead view of the secret U-2 test site, showing the 6,000-foot runway leading onto the southern edge of Groom Dry Lake. Most project personnel called this place "The Ranch," but the official name during the U-2 days was Watertown Strip. (via Mick Roth)

Members of the Skunk Work's Flight Test Group are shown here in front of a Lockheed Jetstar. From left: flight test engineers Ernie Joiner, Jerry Carney, Bob Klinger; test pilots Bob Schumacher and Ray Goudey; and Kelly Johnson. (via Bob Klinger)

These CIA aircraft lined up at The Ranch are carrying the National Advisory Committee for Aeronautics (NACA) logo in a yellow tail band, and a NACA serial number, to support the cover story that the U-2 was a research aircraft. A T-33 trainer used for initial checkout and the Skunk Works' C-47 are visible, left. (Lockheed C88-1447-47)

The CIA recruited the pilots for Project AQUATONE from Strategic Air Command fighter wings that were deactivating. This group at Turner AFB, posed in front of an F-84F circa 1954, includes two future CIA U-2 pilots, Frank Powers (left) and Carl Overstreet (right). (via Sue Powers)

Lumbering C-124 transports were used to move disassembled U-2s into The Ranch for flight test, and out again to the operational detachments. The wings and fuselage were mounted on wheeled carts, and covered for extra security. (Lockheed C88-1447-08)

Colonel Jack Gibbs was the senior military officer assigned to the CIA's Development Project Staff (DPS) from 1956 to 1958. He helped Bissell manage the operational debut of the U-2, and also began the search for the aircraft's replacement. (via Jim Gibbs)

USAF Chief of Staff General Nathan Twining briefs journalists in Moscow, during his visit for the Tushino air show in June 1956. The first U-2 flights over the Soviet Union were postponed until Twining and his delegation had returned to the US. (NARA 155066AC)

Three Bison *bombers fly over Moscow during the Tushino airshow, 23 June 1956. This was the first Soviet long-range jet bomber and at this time, Western intelligence had little idea how many were in service. (NARA 155107AC)*

The unlikely location for the CIA's Photo Intelligence Division (PID) was the top four floors of this building at 5th and K Street in northwest Washington. It was known as the Steuart Building, after the automobile dealership which occupied the ground floor. (CIA)

Art Lundahl was the driving force behind PID, and made sure that the technical advances in collection that the U-2 made possible, were matched by superior photo interpretation and analysis. His presentation skills were frequently employed to brief Presidents and Prime Ministers and others on the results of Project AQUATONE. (CIA)

An A-2 camera rig is winched into the U-2 payload bay, which spanned the entire fuselage immedi-ately aft of the cockpit. This interim configuration of three 24-inch focal length units in fixed mounts was used for the first series of overflight missions. (Lockheed C88-1447-52)

The characteristic "herring-bone" pattern of an S-25 (SA-1) surface-to-air missile site is apparent in this U-2 image, taken near Moscow on the only overflight of the Soviet capital. The notations at the edge of the image read "5 7 56 A2014 TOP SECRET TALENT," indicating the date of the flight, the type of camera prefixed to the mission number, and the classification and codename applied to the product. (CIA)

The S-25 system defending Moscow featured a strange double rotating disc array for missile guidance, which puzzled Western intelligence analysts at first. (Russian Aviation Research Trust)

Fortunately for CIA pilot Carmine Vito, none of the R-113 missiles for the S-25 system were actually in place at the firing sites, when he flew over Moscow on 5 July 1956. (Russian Aviation Research Trust)

Sol'tsy was one of many Soviet bomber bases which were photographed on the first series of U-2 overflights. Photo-interpreters counted 25 Badger *medium-range bombers on this image. However, there was no sign of the long-range* Bisons *or* Bears, *which could have carried nuclear weapons to the U.S. mainland. (CIA)*

Before the U-2 began flying over the Soviet Union in 1956, Western intelligence under-estimated the capability of Soviet early-warning radars like the Token *system shown here. To President Eisenhower's great disappointment, the early flights were detected, and partially tracked.*

Soviet interceptors, like these MiG-17s flown over Moscow during the 1956 airshow, tried to challenge the intruding spyplanes. The U-2 pilots observed their contrails through the driftsight, but with a maximum altitude of only around 50,000 feet, they were not a serious threat.

Previous page, bottom: *Seen here at Wiesbaden in Germany during 1956, Carmine Vito (left) and Marty Knutson (right) were two of the pilots from Detachment A who flew the first missions over "denied territory" in the U-2. (via Jim Wood)*

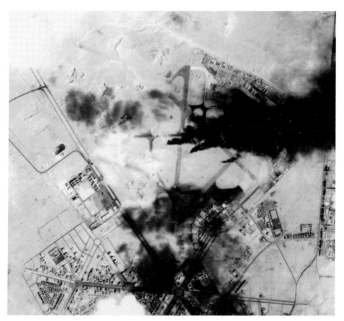

Black smoke rises from Almaza airfield, Cairo, shortly after it was attacked by British and French bombers on 1 November 1956 during the Suez Crisis. The photo was taken by a U-2 from 70,000 feet, during one of the regular flights mounted by the CIA's Detachment B over the Middle East (via Dwayne Day)

The U-2 also carried Signals Intelligence (SIGINT) payloads, and the largest of these occupied the entire payload bay, replacing the camera system, as shown here. Starting in late 1956, such systems were used on flights along the Soviet border which did not penetrate.

Cigar-chomping SAC commander General Curtis LeMay was jealous of the CIA's U-2 operation, and made two attempts to wrest the overflight mission from civilian control. (NARA 151278AC)

This was the first photo of the U-2 to be officially released, but not until February 1957.

Dr Ed Purcell, the Harvard scientist and member of the original Land Panel whose theories of radar deception prompted the attempt to make the U-2 "stealthy." (Harvard University Archives)

Radar absorbent materials were applied to the U-2 during Project Rainbow, in the hope that they would render the aircraft invisible to Soviet early-warning radars. As can be seen, only the lower half of the fuselage was treated. (Lockheed)

In a bizarre further attempt to fool Soviet radars operating in the metric (VHF) wavebands, wires were strung across the aircraft's outer skin at carefully-measured distances, and from booms attached to the wing and tail. (Lockheed)

Bob Sieker, the Skunk Works test pilot who was killed in the crash of the prototype U-2 on 2 April 1957. (Lockheed 05626)

The USAF began receiving its own U-2s at the Ranch in September 1956. 56-6696 was the third one delivered, but this official photo wasn't released until later in 1957, some time after the aircraft had moved to Laughlin AFB, Texas. (Lockheed LA546)

The wreckage of the prototype Article 341 in the Nevada desert. As was often the case when a U-2 crashed after loss of control at high altitude, the aircraft was still substantially complete upon impact. This time, however, a fire broke out and destroyed much of the forward fuselage. (CIA)

In their tight-fitting MC-3 pressure suits, Detachment B pilots and support officers pose for a rare group photo sometime in 1956-57. This group conducted the most intensive series of Soviet over-flights, in August 1957. Top row, from left: Frank Powers, Sammy Snyder, Tom Birkhead, Col Ed Perry (commander), E.K. Jones, Bill McMurry, and Bill Hall. Bottom row, from left: Lt Col Chet Bohart and Lt Col Cy Perkins (operations officers), Buster Edens, Jim Cherbonneaux, and Maj Harry Cordes (navigator). (via Sue Powers)

Exhaustive mission planning was the key to success in any U-2 mission. The objective was to cover the maximum possible number of targets; the choice of camera and operating mode was crucial. Flight lines, turns, and sensor operating instructions were added to topographical maps, which the pilot would take with him. (Lockheed C88-1447-77)

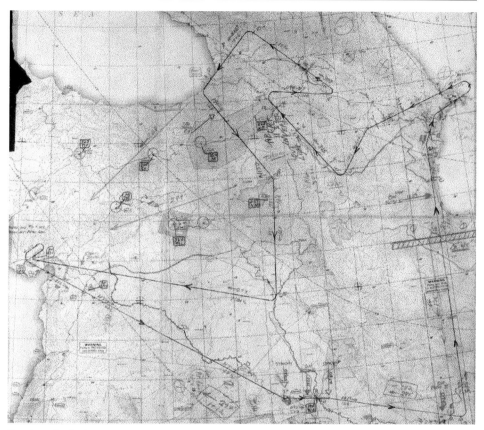

Portion of a flight map issued to Frank Powers for his overflight of Iraq and the southern Caucasus on 20 November 1956. Notations along the indicated flight line show headings, and where to turn cameras on and off. Abort routes are also marked, and Powers used one of these to return to Incirlik, Turkey, when an inverter failed two-thirds through the flight. (CIA)

The Hycon HR73, or "B-camera" was an innovative and superbly engineered system. The lens, shutter, and mirror assembly (bottom right) rotated to seven selectable positions. Two rolls of contra-winding 9 x 18-inch film were fed past the vertical platen from huge magazines (top), each of which housed up to 6,500 feet of the thin, mylar base film.

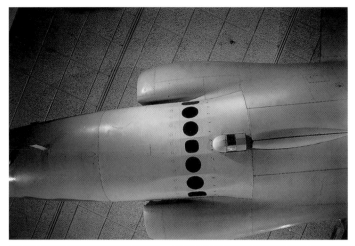

The seven windows of the B-camera hatch are evident here, as well as the extent of the U-2's camera bay. This was designed to take modular payloads, and later became known as the "Q-bay." (author)

The launch site for the R-7 was pinpointed by two U-2 overflights of the desert area east of the Aral Sea in August 1957. To disguise the location, the Soviets codenamed it "Tashkent-50," so the CIA named it Tyuratam after the nearest railway station shown on their wartime maps of the area. (CIA)

The R-7 (SS-6 Sapwood) was the first Soviet ICBM, as well as the launch vehicle for Sputnik and the Luna probes. In the U.S., fears that the huge, liquid-fueled rocket was being deployed in large numbers led to the missile gap controversy. (Russian Aviation Research Trust)

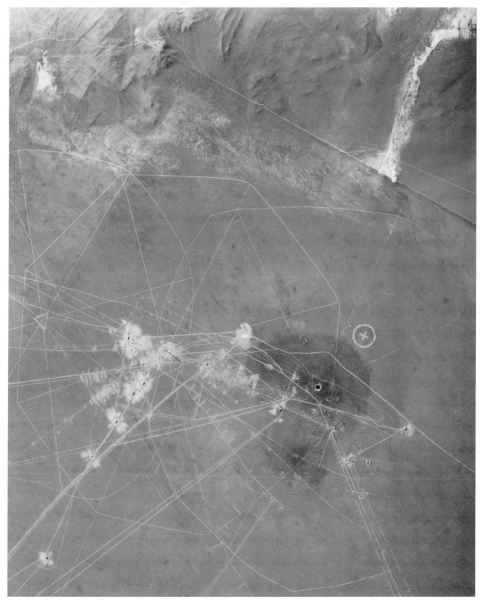

Unknown to pilot Jim Cherbonneaux, the prime target for Mission 4050 on 22 August 1957 was the Soviet nuclear weapons test site at Semipalatinsk. This image shows some of the eight ground zeros and weapons effect test areas that were captured by the B-camera. (CIA)

Four CIA U-2s within the tight but secure confines of Detachment C's hangar at Atsugi in late 1957 or early 1958. The aircraft in the foreground carries the new wing-mounted slipper tanks for extended range. It is also a "Gray Ghost"—an aircraft with radar-absorbent covering. (Lockheed PR-1033)

An overflight of Soviet Northern Fleet bases on 13 October 1957 by Detachment A pilot Hervey Stockman yielded good imagery of submarines, cruisers, and their support facilities onshore. (CIA via Rolf Tamnes)

Trailer life at Detachment B, Incirlik. Time passed slowly for the men assigned to this remote base in southeastern Turkey, enlivened mainly by marathon card games, barbecues, and large amounts of alcohol. (via Bob Murphy)

Another party scene at Det B, with the Agency's executive officer John Parangosky at the center of attention. Onlookers include unit commander Col Stan Beerli (extreme left) and pilot Al Rand (extreme right). (via Bob Murphy)

Before it closed in November 1957, the CIA's Detachment A staged this unique formation flight over Wiesbaden, Germany.

Colonel Jack Nole commanded the 4028th SRS, the USAF's U-2 squadron. On 26 September 1957 he made the highest-recorded escape from an aircraft when bailing out of a crippled U-2 over Texas.

One of the six USAF aircraft with specially-modified noses for nuclear fallout sampling. The nose intake door is open, and also the side panels which provided access to the filter papers. This photo was taken at Ramey AFB, PR, in late 1957 during the 4080th wing's first deployment on Operation Crowflight. (USAF via Tony Bevacqua)

An airman removes one of the filter papers after a U-2 sampling mission. The minute particles of long-term fallout captured on their cotton fibers were carefully analyzed by government and contract laboratories. (NARA 164464AC)

The large airscoop of the F-2 foil system, which was used to collect nuclear weapons debris for intelligence analysis, is visible on this USAF U-2A, forward of the main undercarriage. The P-2 gaseous sampling system was usually carried at the same time, both being housed in the payload bay. The Crowflight badge adopted by the 4080th SRW is on the tail. (NARA K-KE 47602)

The polished metal finish on the CIA's U-2s gave way to a very dark blue in 1958, since this made them less visible at high altitudes to opposition pilots attempting to intercept. This U-2A from Detachment C is landing at Atsugi.

Four British pilots were sent for training at Laughlin AFB in April 1958. From left: Sqn Ldr Chris Walker and Flt Lts John MacArthur, Mike Bradley, and Dave Dowling. Walker was killed in a training accident, but the other three graduated and moved on to the CIA's Detachment B in Turkey. (via Pat Halloran)

Landing the long-winged U-2 was always a challenge. For a time, the 4080th wing used the tail parachute, as shown here. (NARA K-KE 12287)

tests kicked off on 28 May, producing a fireball 900 feet wide. The AEC preferred to set off the tests when the wind would blow the fallout northeast, away from Las Vegas and the populated areas of California. The Groom Lake area was therefore usually directly under the fallout path, and had to be evacuated each time.

With the PLUMBBOB series scheduled to last for many months, a decision to abandon the Ranch was reluctantly taken. The CIA would establish a test unit to continue development work on the U-2 at Edwards AFB, in a remote location at the northern edge of the Rogers dry lake named North Base. It would be supported by Lockheed flight test and maintenance personnel. The growing fleet of USAF aircraft would move out to the permanent base which SAC had chosen for the U-2. This was another remote location at Laughlin AFB outside Del Rio, on the border between West Texas and Mexico.

The U-2 was going semi-public, but the USAF was still wrestling over the security classification to apply to its strange-looking new jet. The 4080th wing, which moved to Laughlin in April where it displaced a pilot training wing, was told that "a positive attempt will be made to release very little information."[6] SAC devised cover names for the U-2's photographic and SIGINT payloads, in an attempt to associate them with the "weather research" mission. Cameras became "Nepho," short for nephograph: an instrument for photographing clouds. SIGINT systems became "Sferics," a shortform of "atmospherics."

No Big Secret

In reality, the true mission of the U-2 was already obvious to anyone with a modicum of intelligence skill. Within a month of Det A's arrival at Lakenheath in May 1956, the UK's Flight magazine ran two sarcastic editorials scoffing at the cover story and the excessive security surrounding the deployment. "Very high weather we're having lately!" remarked another British magazine. Soon the aircraft spotters which gathered around the perimeter of British military bases were taking snatched photographs of U-2s on approach to Lakenheath—and sending them to the aeronautical press. In February 1957, the U.S. government finally released the first official photograph of the U-2—an aircraft in NACA markings. Soon after Det C began operations from Atsugi, a Japanese photographer snapped one of the aircraft on approach, and the result was published in Koku Joho (Aviation News) magazine in May 1957. Security officers from the base allegedly searched his home while he was out.[7]

Then, on 30 May 1957, the CIA's cover story was well and truly blown. The London Daily Express reported that "Lockheed U-2 high-altitude aircraft of the U.S. Air Force have been flying at 65,000 feet, out of reach of Soviet interceptors, mapping large areas behind the Iron Curtain with revolutionary new aerial cameras. They are making mathematically precise maps essential to bombardment with missile weapons."

The true nature of U-2 operations also leaked into the U.S. press, but not into the major newspapers, since top editors were persuaded by the government to suppress the story in the national interest. The security for the U-2 project was run by a special CIA team in Project HQ headed by Walt Lloyd. Many workers within the project thought the security team was over-zealous, to say the least. But the rigid policy of admitting no connection between the U-2 and airborne reconnaissance continued. Maybe the security team calculated that Soviet intelligence relied upon the U.S. press for all its U-2 information, and never monitored the British or Japanese press!

Death in Texas

On 10 June 1957, the CIA and Lockheed moved from the Ranch to the North Base at Edwards. The next day, Colonel Jack Nole led a formation of four USAF U-2As from Nevada to Laughlin AFB. A further six aircraft arrived at the Texas base over the next two days, to continue the extensive training program which SAC had devised. The 4080th wing had been allocated primary responsibility for the sampling mission, but would also train for photo and SIGINT missions. The 4028th Strategic Reconnaissance Squadron (SRS) was activated to operate the U-2s, with Nole as the commander. The U-2 squadron would operate alongside the RB-57Ds of the 4025th SRS, which had moved from Turner AFB to Laughlin a few months earlier.

After just eighteen days at Laughlin, SAC's "secret" U-2 operation was compromised. The first pilot to be checked out at Laughlin decided to take a trip downtown—*with* his airplane! Lieutenant Ford Lowcock had set up house on the outskirts of Del Rio near to the small civil airport. Just before nine in the morning on Friday 28 June, on only his second checkout ride in the U-2, Lowcock flew across town at 1,500 feet. Apparently, he wanted to stage a private airshow for his wife and two small sons. Lowcock made two descending turns to the right over the house, but he was unable to bring the wings back level. The aircraft spiraled into a ceniza-covered hill next to the airport, and flipped upside-down. There was no fire, but the 28-year old Lowcock was knocked unconscious, and died from asphyxiation. He hadn't realized that, with the U-2's ailerons in gust control, their effectiveness in roll was much reduced.[8]

Only two hours later, First Lieutenant Leo Smith took off from Laughlin for a high-altitude training flight and he, too, did not return. The plane crashed about 45 minutes later some 30 miles north of Abilene, Texas. The accident investigation determined that an autopilot or trim malfunction might have occurred, leading to loss of control. Another possible cause was fuel imbalance in the wings (a contributory factor in the Lowcock accident). Oxygen problems were also on the list of possible causes. Smith had evidently tried to escape, but—just like Sieker two months earlier—he had been pinned back by g-forces, and the bulky pressure suit hadn't helped. After these two accidents, the USAF agreed with Kelly Johnson that an ejection seat should be provided, and work began to adapt Lockheed's own light-weight T2V seat. But there was no decision to adopt the dual oxygen regulator, another safety item which Lockheed had recommended.

"Dirty Birds" Deploy

After the White House meeting in early May 1957 which gave preliminary approval for more Soviet overflights, Project HQ decided to deploy the U-2s with the radar-evading modifications, even though flight tests were not yet complete. In June, one modified U-2 was sent to each of the CIA detachments. They were not enthusiastically received. "Part of my big paycheck was compensation for high-risk missions in a semi-experimental airplane, but I had never before risked flying an airplane wired like a guitar!" recalled Det B pilot Jim Cherbonneaux.[9] The Skunk Works engineers had named the modified aircraft "Dirty Birds," in an obvious comparison with the aerodynamically very "clean" standard U-2. The detachments called them "Covered Wagons."

Jim Cherbonneaux flew the first operational test of a Covered Wagon on 21 July 1957. This was a six-hour flight along the Soviet border at 12 miles distance, using the System 4 SIGINT package in an attempt to determine whether the modified aircraft could fool the air defense radars. Cherbonneaux had already flown a similar SIGINT mission along the Soviet border three months earlier, in a conventional U-2. This earlier flight was designed to collect the latest Soviet radar frequencies for use in the Rainbow project, and also to determine whether the radar defenses were weaker to the east. Cherbonneaux had flown all the way across Iran to Afghanistan above a solid undercast, failing all the while to pick up any useful signal for the radio compass, which could help him fix his position. Reversing course, but still relying on dead reckoning, he strayed across the border near Baku. Cherbonneaux hadn't realized this until he saw the contrails of Soviet interceptors heading his way. The UI-2 pilot beat a hasty retreat to the south.[10]

This time, though, the weather was fine as Cherbonneaux flew the Covered Wagon all along the Soviet Black Sea coast and back. Occasionally, his flight plan called for the aircraft to fly towards the coast rather than parallel, to deliberately provoke a radar response. The NSA's ground stations in Turkey would also be listening for a Soviet reaction. The results of the mission were inconclusive. There had been a complication during the flight, when the engine bleed valves had opened at random a few times, causing each time a rapid 4-5,000 foot descent through loss of power. Cherbonneaux flew another long test on 30 July. On both flights, System 4 recorded some alerting of Soviet radars, especially when the aircraft was flying directly towards or away from the radar sites. Upon further study, it was realized that most of the radar returns were coming from the U-2's cockpit, engine intakes, and tailpipe, which could not be covered with absorber.

The results were good enough for Bissell, though, as he pressed the case for more overflights. The CIA had contacted the President of Pakistan, who had given permission for U-2 overflights from his country. That was a double bonus. Soviet air defenses were weaker in that area than anywhere else. And many of the highest-priority targets were deep inside the USSR, from the Urals to Siberia. Even with its long range, the U-2 couldn't reach them from anywhere else.

At Project HQ there was a flurry of activity as multiple mission plans were drawn up. Each one was planned around one or two of the highest priority targets on the ARC list. Then as many lower-priority targets as possible were added. At Incirlik, Det B commander Colonel Ed Perry prepared a deployment plan. In late July, C-124 transports lifted all the required fuel and support gear to the airbase at Lahore. It was the CIA's second choice, but the preferred airbase further north at Peshawar was closed for runway repairs. Two U-2s were ferried in, and hidden inside a hangar which had been borrowed from the Pakistan air force. Operation SOFT TOUCH was about to begin. Early on 5 August 1957, two U-2s flew off to the north, but one had to abort and return. The other, flown by Eugene "Buster" Edens, continued across the border and began the first deep penetration of Soviet airspace for over a year. Mission 4035 was searching for the very highest-priority target—a suspected new Soviet long-range missile launch base.

Soviet Missiles - Big and Small

By the middle of 1957, the USSR had been testing its new series of ballistic missiles from the Kapustin Yar test range south of Stalingrad for four years. Starting with the short-range R-11 (which became known in the West as the SS-1 *Scud*), the design team led by Sergei Korolev had progressed to the R-5M nuclear-tipped missile with a range of 750 miles. The R-5M was cleared for service in 1956, and the first launch sites chosen by the Soviet Army in the Baltic States were under the flightpaths of the early U-2 overflights. It was designated SS-3 and nicknamed *Shyster* by Western intelligence. But the R-5M was still a cryogenic-fueled missile which derived from the wartime German V-2 design.[11]

Since mid-June 1955, the U.S. had been monitoring launches from Kapustin Yar with the AN/FPS-17 ground radar at Diyarbakir in Turkey. Designed by General Electric for this specific purpose, it was the world's largest and most powerful operational radar.[12] Other ground stations in Turkey monitored Soviet down-range communications whenever a launch was planned, and attempted to intercept telemetry from the missile in flight. On 22 June 1957, the big radar observed a missile launch from Kapustin Yar which impacted much further east in the remote Kazakhstan desert than usual. Further tests of this 1,100-mile range missile followed in quick succession. Meanwhile, SIGINT and other intelligence suggested that a new test range was being built to handle intercontinental-range missile firings, from a launch site somewhere east of the Aral Sea and an impact area on the Kamchatka Peninsula, 4,000 miles away.

The CIA had already made two attempts to photograph the Kamchatka area with the U-2. On 8 June, Det C pilot Jim Barnes flew north out of Atsugi in a first attempt to overfly the suspected impact area. But the entire Kamchatka peninsula was covered in cloud. Barnes therefore stayed offshore, and flew the aircraft on to the planned destination at Eielson AFB, Alaska. On 18 June, the weather forecast for Kamchatka was better, so Al Rand took off for a second attempt. He completed the mission as briefed, returning to Eielson. However, there was still a lot of cloud, and a camera malfunction ruined every third frame. Nevertheless, the photo-interpreters were able to discern some preparations for missile tests around Klyuchi.[13]

It was clear that the Soviet missile program was shifting into higher gear. In fact, the brilliant but abrasive Korolev had a new rival: his former protege Mikhail Yangel. Korolev's team at the Kaliningrad design bureau had been working on his major ambition, a huge liquid oxygen-fueled booster capable of thrusting vehicles into space, as well as lofting nuclear weapons into the American heartland. But the Soviet army wanted a practical intermediate-range missile, too—one that was mobile and used storable fuels. Yangel was allowed to set up a new design bureau at Dnepropetrovsk in the Ukraine to meet this need. It was Yangel's R-12 missile that had flown first, and this was the unusual test flight from Kapustin Yar which the Diyarbakir radar had observed. But development by Korolev and his team of an intercontinental missile was not far behind. It was designated the R-7, but Korolyev called it the Semyorka ("Little Old Seven"). This missile needed no less than 400 metric tons of thrust for liftoff; a much more accurate inertial guidance system; and a dedicated transport and launch infrastructure.[14]

In 1955, construction of a launch site for the R-7 had begun in the Bet Pak Dala Desert near the railway line which followed the Syrdar'ya river from the Aral Sea to Tashkent. Even when it was disassembled, rail was the only practical means of transporting the R-7 over long distances. In common with many other secret Soviet weapons sites, the new launch site was given a "postbox" number in the nearest large city, to hide the true location. In early 1957, assembly of the first R-7 began inside a huge hangar at postbox "Tashkent-50." The four strap-on boosters were carefully mated to the central core stage. Then it was moved out by rail to the Tyulpan (Tulip) launch structure, so-called because the R-7 sat in its center, stabilized on all sides by four enormous "petals" containing the work gantries and other structures needed to fuel and test the missile.[15]

After two failed attempts, the first R-7 was launched from "Tashkent-50" on 15 May 1957, but after only 50 seconds it exploded over the test range. On 11 June 1957, another launch also failed. None of this was known to the CIA, although SIGINT intercepts told of heightened preparations along the test range. And so when Buster Edens flew his U-2 into the Kazakh SSR on 5 August, part of the route followed the railway which ran northwest from Tashkent, searching for the new launch site. The B-camera with its option for horizon-to-horizon coverage proved invaluable on this type of mission. When the developed B-camera film reached the PID in Washington a few days later, the huge launchpad was visible on one of the high oblique frames. But it was little more than a speck on the horizon, since the launchpad was located at the end of a spur extending some 15 miles into the desert from the main rail line.[16]

PID's all-source analyst Dino Brugioni turned to the best available map of the area, which was still the 1939 map drawn up by the German Wehrmacht. This map also showed a spur leading north from the station at "Tjuro-Tam." Brugioni deduced that the spur had been built to serve a prewar quarry, which had been converted to serve as the flame pit for the large new Soviet missile. "Tashkent-50" was now identified, and would henceforth be known as Tyuratam in the West.[17] A second U-2 mission was quickly requested, to fly directly over the launch pad.

Targets, Cameras, Routes
Meanwhile, in the cramped hangar at Lahore, Det B stayed on alert for more missions. In addition to the two missile launch sites and their downrange installations, the reconnaissance target list was rich with possible nuclear weapons development sites and military industrial complexes. The furthest of these were nearly 2,000 miles from Lahore, in the Kuznetsk Basin, known as "Stalin's second industrial bastion." Many of the targets were "secret cities," completely closed to foreigners. For instance, there was Krasnoyarsk, where a huge new complex of apartment houses, laboratories, warehouses, and machine shops had been revealed in 1956 by one of the few balloon flights to produce any useful intelligence. There was also Tomsk, where returning German prisoners of war had suggested there was a "secret undergound atomic plant." The CIA arranged for the fur hat which one of these Germans had worn to be analyzed by nuclear scientists. The hat contained traces of the U-235 isotope, suggesting that Tomsk was a separation plant.[18]

Other targets which were listed for the new U-2 missions included uranium concentration plants in the Kirghiz and Tadzhik SSRs; an atomic energy complex and a factory producing fighter-interceptors at Novosibirsk; another aircraft plant in Omsk; and a rocket motor production and test site at Biysk.

Some of the targets could not be located with enough certainty to guarantee that the U-2 would pass overhead, and thus provide the best-quality imagery. This was true of the impact areas for missiles launched from Kapustin Yar, somewhere west of Lake Balkhash, and the nuclear weapons proving ground which was thought to be near Semipalatinsk. In the latter case, the only intelligence available was seismic data, but none of the twenty nuclear tests which had taken place here in the previous eight years could be fixed closer than within 30 miles. Therefore, the mission planners had to orient the flight lines for maximum coverage of the most likely territory, or specify that the U-2 perform some search patterns.

A careful choice of camera type and mode of operation was also vital. The B camera was producing superb results. Photo-interpreters habitually dismissed any images with a scale greater than 1:8000 as marginal to their trade. But when the 36-inch lens of the B produced excellent imagery from 70,000 feet at a scale of 1:33,000, they had to revise their thinking. But the B-camera did have *some* limitations. For instance, if the Mode 1 operation, which used all seven "stop and shoot" positions to give horizon-to-horizon coverage was used, the film ran out after five hours of operation. Moreover, the high-oblique imagery obtained at the furthest of the seven positions was more difficult to interpret, even though AUTOMAT had the latest rectifiers which could correct the image to constant scale. The main reason for covering horizon-horizon was to use the geographic features so captured, such as mountains or lakes, to orient the more vertical images.

If the target location was known, the B-camera's Mode 2 was a better option. This eliminated the high and medium oblique stops; allowed eight hours continuous operation; and the resolution looking straight down was the best possible—up to 100 lines per millimeter. But this mode photographed a swath of territory which was less than ten miles wide—so there was not much room for error in planning the flight track, or flying it!

Another peculiarity of the "B" was the contra-winding film, with matching halves of each image exposed on each of the two 9 x 18-inch rolls. Inevitably, a small strip of terrain down the middle of the complete 18 x 18 image could not be captured on film. This wasn't properly appreciated until the first overflight mission with the B-camera in September 1956. An eastern European airfield was overflown to determine whether a runway had now been built there. But the U-2 flew directly overhead so that the middle of the airfield "fell between" the two halves of the image. The runway analysts were none the wiser.

Sometimes, the more conventional A-2 trimetrogen camera configuration was preferred, because it photographed a 40-mile wide swath, and there was a small overlap between the left, center, and right images. Here, though, the trade-off was resolution, since the A-2 used three 24-inch focal length lenses instead of the 36-inch lens in the B-camera.[19]

When all the target and camera choices had been made, the mission plan was transmitted to the Det, where the pilot maps were drawn up. The flight lines on the map were drawn in color: red for portions of the flight where the course had to be followed exactly so that a particular objective was covered, blue for less important sections, brown for emergency abort routes back to base. The straight flight lines were linked by "ten-cent turns," so-called because they could be drawn on the map with a dime. Notations on the map instructed the pilot when to turn the camera and SIGINT systems on and off, or select different operating modes for them.

A second overflight was launched from Lahore on 12 August, and then a pause. There was very little indication of Soviet tracking, and no protest note. A third aircraft was flown in from Incirlik and squeezed into the hangar. Then on 21 and 22 August, as fine weather prevailed across the entire southern USSR, two more missions were flown each day. It was an enormous effort, and not only for the maintenance crews. Although the mission plans were devised in Project HQ and transmitted to the Det, the waypoints still had to be transcribed onto the pilot's maps, and the fuel curves plotted. Under great pressure, the detachment's operations officer, Major Harry Cordes, did most of this work in the hangar at Lahore. On the 22nd, he actually had to draw up maps for three overflight missions which took off in quick succession.

The third of these missions did not penetrate the USSR, but flew over Tibet instead. This mission, flown by Bill Hall, was designed to map virtually the whole of this remote country. The CIA had just begun supporting a group of Tibetan exiles in their attempts to expel the communist Chinese who had invaded their homeland in 1950. The U-2 photography helped to identify suitable infiltration routes and paratroop drop zones. These airdrops began two months later, in December 1957.[20]

Five other flights in the SOFT TOUCH series also flew over China, but only its western extremes as they headed to or from their main objectives deep into Siberia. But in the gin-clear skies, pilots enjoyed spectacular views of K-2 and other mountains in the Karakorum range, as well as the Xinjiang desert to their north.

Lift-Off!
The last overflight from Lahore was launched on 28 August, when E.K. Jones flew the return trip to Tyuratam, which the missile analysts had urgently requested. Exactly a week earlier, the huge R-7 had finally achieved a successful lift-off there. It was Korolev's last chance. Another failure, and Premier Khrushchev had threatened to close down his design bureau. In fact, the 21 August flight of the R-7 was not a complete success. The re-entry stage carrying a mock-up of the nuclear warhead reached the specified range over Kamchatka, but disintegrated into a few thousand pieces at 30,000 feet.[21] Nevertheless, on 26 August Moscow announced the successful flight, adding pointedly that "it is now possible to send missiles to any part of the world." But the announcement still did not identify the launch site. Eventually, the USSR misleadingly named it Baikonur, after a town which was located fully 200 miles to the northeast. But on the 28 August flight, Jones obtained excellent vertical photography of the Tyuratam site, which showed just one launch pad. Mission 4058 only just missed photographing the rollout of the next R-7, which was successfully launched on 7 September in the presence of Khrushchev himself.[22]

Desperate to get some idea of how accurate this missile could be, the CIA tried again for U-2 coverage of the impact area on Kamchatka. The fourth "Dirty Bird" was ferried from Edwards to Eielson, from where Det C pilot Barry Baker flew Mission 6008 and successfully photographed the Klyuchi area on 16 September. Eighteen days later, yet another R-7 was launched from Tyuratam. But this particular test flight had a different, and highly public, purpose. The guidance system and dummy warhead in the nosecone had been replaced by a small satellite! The firing sequence of R-7's core engine and strap-on boosters was modified to push the missile higher into the stratosphere. The successful launch into earth orbit of Sputnik 1 by the R-7 on 4 October 1957 was a propaganda and technical coup for the USSR. The world sat up and took notice.

Korolyev's "Little Old Seven" had proved itself in spectacular fashion, and the "missile gap" controversy began. But the U-2 was providing some solid information about Soviet missile flight tests. It had soared over Tyuratam and Klyuchi, and in another overflight which wrapped up the SOFT TOUCH series on 10 September, Bill Hall flew the aircraft over Kapustin Yar for the first time. One of Yangel's new R-12 missiles was photographed on the launch pad.

Moreover, the U-2 had confirmed the site where nuclear warheads for the new missiles were tested. On one of the three flights launched from Lahore on 22 August, Det B pilot Jim Cherbonneaux had been flying over Soviet territory for about three hours when he noticed some familiar-looking landmarks beneath him. There were large circular areas of ground which had been cleared and graded, with paved support roads leading outwards towards distant block houses. It was just like flying over the AEC proving grounds in Nevada from the Ranch, the pilot recalled! When the plan for this mission was transmitted from Washington, Det B had not been told that this flight was designed to find the Soviet nuclear test site. In fact, CIA nuclear analyst Henry Lowenhaupt had calculated the average epicenters of the five largest nuclear detonations in the area. He requested a flight track which passed over this spot, which was 70 miles due west of the inhabited area at Semipalatinsk. So the mission planners at Project HQ routed the flight west-to-east between Karaganda and Semipalatinsk, for maximum coverage. And on U-2 Mission 4050, they hit paydirt!

Cherbonneaux adjusted course slightly to fly over the centre of one of the cleared areas. He brought the driftsight up to full x4 magnification, and scanned around the site. There was a large tower at the centre, which was completely deserted. But at a large block house in the far distance, a number of vehicles were parked. Suddenly, the pilot imagined that the tower might be holding a nuclear weapon, which was about to be detonated! Even at 400 knots TAS, it seemed an eternity before the U-2 was clear of the area. Cherbonneaux set course for his next targets in the Omsk area, and breathed a sigh of relief.

When Cherbonneaux returned to Lahore and told his debriefers about the nuclear test site, they didn't believe him. But within 24 hours, Project HQ confirmed that a half-megaton nuclear test had taken place there about five hours after the U-2 flew overhead. When the film was analyzed, it showed a bomber aircraft at Semipalatinsk airfield. This was subsequently determined to be the carrier for the 22 August explosion, which was actually an air-dropped nuclear weapon. But the tower which Cherbonneaux had observed did indeed also contain a nuclear weapon, although this relatively low-yield device was not detonated until 13 September.[23]

Analysis

It took the photo interpreters at PID many months to identify all the interesting nuclear sites from Operation SOFT TOUCH. Some uranium mines were captured on film, but not discovered for a whole year. The "secret underground atomic plant" at Tomsk turned out to be a huge complex covering 40 square miles: not only a U-235 gaseous diffusion separation plant, as expected, but also one plutonium-producing nuclear reactor in operation; two more nuclear reactors under construction; and a plutonium chemical separation facility. But amidst the intelligence bonanza, there were inevitably some disappointments. On the 12 August mission, only 125 exposures were made before the camera malfunctioned.[24] Unfortunately, the Krasnoyarsk area was covered in heavy cloud (it was indeed another secret nuclear city, where uranium enrichment and plutonium production was accomplished).[25]

It took the CIA's missile and nuclear analysts even longer to extract maximum intelligence value from this series of flights. Often, collateral information was acquired at a later date, which prompted a second, third, or even fourth look at the original imagery. The analysts used every possible means to try and understand what they were seeing on the U-2 film. Scale models were built of important sites, such as Tyuratam. Project HQ was asked to fly U-2 photo missions over the Nevada nuclear test site, in order to compare the imagery with that from Semipalatinsk, in case some ground features were common to both sites.

Lessons about the Soviet nuclear program—such as the plutonium production rate—were still being learned from the Tomsk photography five years later. Analysts eventually realized that they could determine the water flow through the Tomsk reactor and its rise in temperature. From that, they could deduce the power output of the reactor and its annual production of plutonium in kilograms. By comparing the techniques used in U.S. nuclear reactors, factoring in the limited technical data available from open Soviet technical literature, and closely studying a cinefilm of "a new Siberian atomic power plant" that Soviet scientists revealed at the 1958 Geneva Conference on the peaceful uses of nuclear energy, the CIA analysts were eventually able to make confident estimates of the USSR's total plutonium production.[26]

PVO tracking

The USSR detected only two of the eight overflights staged from Pakistan during August. From the CIA's perspective, this was a big improvement, but it wasn't thanks to the radar-evading Dirty Birds. They had only been used on a few of the flights, and although Barry Baker's trip to Klyuchi in September had been flown using one of these aircraft, it had been detected and trailed by five MiGs. Worse still, Baker had been unable to fly higher than 59,000 feet in the Dirty Bird, partly because of its extra weight and partly because of warmer temperatures than had been expected in the upper atmosphere. Looking down through the driftsight, Baker could even make out the bonedome of one Soviet pilot as his interceptor shadowed the U-2 only a few thousand feet below.

In reality, it was the lack of Soviet early-warning radar coverage opposite Pakistan, and further east along the mountainous border with China, which had caused most flights to go undetected. This Soviet deficiency had been exploited to the full in Operation SOFT TOUCH. Some of the flights had even been routed to the northeast via Chinese territory before entering or after leaving Soviet airspace. But on Bill Hall's flight to Kapustin Yar, which was launched and recovered out of det B's home base at Incirlik, PVO had no trouble in tracking the U-2 thanks to their complete radar coverage in the southern Caucasus. Hall's Mission 4059 had been planned to fly even further north towards Moscow, but the pilot cut it short when he saw so many interceptors rising against him. The PVO launched 18 of them that day. Instead, Hall turned south to cross the Ukraine. When the flight reached Kiev, anti-aircraft artillery fired at it. Of course, the shells exploded harmlessly thousands of feet below. But the barrage attracted plenty of attention in Kiev itself, causing rumors to spread that the city had come under an air attack. When he heard of this, Soviet Premier Khrushchev castigated the anti-aircraft units for their useless action.[27]

Col-Gen Yevgeniy Savitsky, the man running the PVO's fighter regiments, was unwilling to concede that his prized interceptors could be out-flown. After one of the flights from Lahore had been detected close to the border, a MiG-19 had been scrambled from Andizhan. From his maximum altitude of 55,000 feet, the pilot reported that he could see the cruciform configuration of an aircraft approximately 10,000 feet above him. Savitsky traveled from PVO headquarters in Moscow to interview the pilot. After analyzing all the data, Savitsky pronounced that it was impossible. The pilot was transferred.[28]

What did the USSR really know about the U-2? After the first series of U-2 overflights, the great and the good of the Soviet aerospace industry was summoned by Premier Khrushchev...Pavel Grushin, Artyom Mikoyan, Pavel Sukhoi, and Andrei Tupolev. Khrushchev asked them what kind of plane this new intruder could be, whether it was capable of carrying a nuclear weapon, and how they intended to deal with it. They all agreed that it wasn't a bomber, but only Tupolev was willing to credit the aircraft's true nature and capability. He designed bombers rather than fighters, and therefore had nothing to lose. He sketched a probable long-duration, high-altitude aircraft for Khrushchev: it had long wings and a single engine. It must be a very light structure, Tupolev surmised. Too light to carry a nuclear bomb, he assured Khrushchev. Tupolev's assessment was not taken seriously by the others. They insisted that the American spyplane had to be a twin-engine design, possibly a modified bomber. Moreover, they were only to willing to credit it with the capability of maintaining very high altitude for a short duration. For the rest of the flight, they believed that it must surely descend, and thus fall prey to the MiGs.[29] Even so, Mikoyan and Sukhoi both set to work on experimental modifications to their prized fighter designs, in an attempt to reach higher altitudes.

On 28 August 1957, the British publicly staked a new claim to the altitude record when the twin-engined Canberra B.2 flew to 70,310 feet. But this was no ordinary Canberra; it was boosted by a Napier Double Scorpion rocket motor, and could not stay at this high altitude for very long. Meanwhile, the official data which had been released in the U.S. on the single-engine U-2, gave its ceiling as only 55,000 feet.

The CIA conspired with Lockheed to convince the Soviets that 55,000 feet really was the maximum. Ernie Joiner and his flight test engineers, Bob Klinger and Mel Yoshi, were instructed to create a fake U-2 flight manual. They sketched out a much heavier airframe with more conventional maneuvering load factors, and the same airspeed limits as the T-33. Graphs were invented showing a much poorer rate of climb to the 55,000 maximum altitude. They even included cockpit photos in the manual, showing instruments whose markings had been altered to correspond with the fake data. The Skunk Works engineers included genuine data on the U-2's weather reconnaissance package—but not on any of the other payloads. Three or four copies of the fake manual were artificially aged with dirt, grease, coffee stains, and cigarette burns. Kelly Johnson turned them over to the Agency's clandestine people in DDP, which presumably "planted" them in the appropriate Soviet channels.[30]

It is not known whether the manuals reached the intended recipients.[31] In any case, the policy-makers in Moscow weren't placing all their hopes on Savitsky's fighters, in their attempts to intercept high-flying aircraft. Within the PVO, development resources were switched to surface-to-air missiles (SAMs). During Operation SOFT TOUCH, U-2s had flown over the test range which stretched 8,000 nautical square miles to the west of Lake Balkhash. One mission had photographed a well-planned, modern community of some 20,000 people on the western shore of the lake. The CIA's analysts determined that this was an area for test-launching SAMs, as well as an impact area for medium-range offensive missiles. They named it Saryshagan.

Enter the S-75
In fact, the PVO had its own proving grounds within the Saryshagan complex. It was testing the new S-75 Dvina surface-to-air missile system here. After the short-

comings of the S-25 Berkut system around Moscow had been realized, the Soviet leadership charged the Lavochkin design bureau, led by Petr Grushin, with leading the development of an all-new air defense system. It would have better high altitude performance and a more compact fire control system. The S-25 used a single-stage, liquid-fueled rocket. Grushin's new missile had a solid fuel booster below the kerosene second stage, and was placed on a single-rail launcher. The S-75 command guidance radar used the track-while-scan technique, which had been patented by a U.S. scientist at the Bell Laboratories in 1954. This radar was initially nicknamed *Fruit Set* by Western intelligence, but later it was renamed *Fan Song*, due to the fan-like sweep of its beam and the chirping sound of its electronic signal. With the missile on a semi-trailer, and the radar mounted on a four-wheel trailer, the S-75 was semi-mobile. A newly-developed VHF radar designated P-12 (Western nickname *Spoon Rest*) provided initial target acquisition for the system at a range of 125 miles, with its Yagi-type antenna mounted on the roof of a box-bodied mobile trailer. Each S-75 regiment had one P-12, and commanded three or more batteries which were spaced some miles apart. Each battery had a Fan Song and six missile launchers, and could salvo-fire three of the missiles at six-second intervals.[32]

The missile for the Dvina system was designated V-750[33] and was shown in public for the first time in Red Square on 7 November 1957. Within months, the first operational systems were being deployed around Moscow, Leningrad, and Baku.[34] The S-75 missile system was given the designation SA-2 and nicknamed *Guideline* by the West. Eventually, it would prove to be the U-2's nemesis.

Notes:

[1] Jack Gibbs remarks, relayed to author through Jim Gibbs
[2] Johnson log, 16-17 August 1956
[3] Richard N. Johnson, "Radar Absorbent Materiel: A Passive Role in an Active Scenario", The International Countermeasures Handbook, 11th Edition 1986, EW Communications Inc, Palo Alto, CA
[4] Robertson, Klinger interviews
[5] Matye, Setter interviews
[6] History of the 4080th Strategic Reconnaissance Wing, filed at AF Historical Research Agency, Maxwell AFB, April-May 1957, hereafter 4080th history
[7] "Open letter on 'Black Jet' Relayed to US commanders", Mainichi, 1 December 1959
[8] accident report, 4080th history, October 1957
[9] as quoted in Rich and Janos book, p154
[10] the unintentional overflight took place on 18 March 1957, but the USSR did not make a formal protest
[11] Stephen Zaloga, "Target America", Presidio Press, 1993
[12] Stanley Zabetakis and John Peterson, "The Diyarbakir Radar", CIA Studies in Intelligence, Fall 1964, declassified 1996
[13] Pedlow/Welzenbach, p134; interview with Jim Barnes
[14] Zaloga, "Target America"
[15] ditto
[16] the author has not been able to confirm the version of the first U-2 flight over Tyuratam as related by Bissell book, p119, namely that the pilot saw the launch site off his nose, and altered course to fly directly overhead. Bissell is certainly wrong to suggest that the existence of the launch site "had not even been suspected."
[17] Dino Brugioni, "The Tyuratam Enigma", Air Force Magazine, March 1984
[18] Henry Lowenhaupt, "Mission to Birch Woods", CIA Studies in Intelligence, Fall 1968, declassified 1997
[19] Brugioni book, Crimmins interview. In a test flight over the North Island Naval Air Station shortly before it was cleared for service, the B-camera was able to resolve the white lines painted on the tarmac in the parking lots. President Eisenhower used this photograph to inform the public of the U-2's capability, in a television broadcast shortly after CIA pilot Powers was shot down in May 1960. He told viewers that the aircraft was 13 miles high when the photo was taken, but the parking lines were only four inches wide!
[20] Tsering Shakya, "The Dragon in the Land of the Snows", Pimlico Books, UK 1999; William M. Leary, "Secret Mission to Tibet" in Air & Space magazine, December 1997/January 1998
[21] Timothy Varfolomyev, "Soviet Rocktry that Conquered Space", Spaceflight, Vol 37, August 1995, p261
[22] According to Varfolomyev, see previous footnote, this launch may have been an unsuccessful first attempt to launch the Sputnik satellite, rather than an ICBM test
[23] Lowenhaupt article
[24] Pedlow/Welzenbach, p135
[25] Lowenhaupt article
[26] Henry Lowenhaupt, "Ravelling Russia's Reactors", CIA Studies in Intelligence, Vol 16, Fall 1972
[27] Sergei Khrushchev interview
[28] Votinsev article, part 2, in Voyenno-Istoricheskiy Zhurnal, No 9, 1993
[29] remarks by Col Alex Orlov at U-2 conference, Washington DC, September 1998
[30] Bob Klinger notes
[31] There is one tantalizing clue that the fake manual may have served its purpose. Pyotr Popov, a Lt Col in the GRU who had been spying for the CIA since 1952, reported to his US handler in April 1958 that he overheard a drunken Colonel boasting that the KGB had "many technical details about the new high-altitude spycraft." Of course, the KGB might simply have aggregated the considerable amount of information about the U-2, which had been published in the Western press by now! (The USSR discovered Popov's treachery later in 1958 and executed him the following year).

[32] Stephen Zaloga, "Soviet Air Defense Missiles" However, the author understands that the overall Dvina system designation (eg radar plus missiles) is S-75, and not V-75 as stated by Zaloga and others.

[33] A reliable Russian source told the author that the missile designation is actually D-20, but V-750 is used hereafter because it is so widely accepted

[34] According to Zaloga, p37, a prototype S-75 system was deployed to the Kala airport at Baku as early as late 1956. It may therefore have been overflown by U-2 Mission 4016 flown by Frank Powers on 20 November that year, which routed directly over Baku.

5

Sampling the Stratosphere

So far in 1957, all the significant action in Project AQUATONE had taken place in Detachments B and C. There had been virtually no operational missions for Det A at Giebelstadt. Morale and discipline had suffered accordingly; the pilots had little to do except fly the Det's two T-33s and single L-20 Beaver (for additional "taildragger" landing practice). After a series of incidents which potentially breached security, Det A commander Col Fred McCoy was replaced by Col Hack Mixson from Project HQ. In the fall of 1957, Project HQ made plans to close Det A. The contracts of the "sheep-dipped" U-2 pilots were coming up for renewal; it was decided to maintain Dets B and C, but reshuffle the personnel.

At the same time, however, a requirement arose to monitor Soviet naval exercises in the Barents Sea. The U.S. Navy—which had representatives on the ARC and in the PID—was keen to discover more about Soviet warship development, and particularly the nuclear-powered submarines. Conventional U.S. and British reconnaissance aircraft had been probing the area for years, and the U.S. Navy regularly sent its own submarines to the area, where they would clandestinely monitor Soviet fleet movements. On one such sortie in 1956, a strange modification had been spotted on one Soviet diesel submarine at Severomorsk. U.S. naval intelligence analysts were still pondering its significance. In fact, it was the prototype Soviet ballistic missile-launching submarine.[1]

Det A was alerted for some flights to the north. It was a long way from Giebelstadt to Murmansk and back, but the Det had just received its first U-2 with the GIANT STRIDE modification. This added a pair of external "slipper" fuel tanks to the wings, each carrying 100 gallons. The effect was to restore the range of the standard U-2A model, which had dipped below 4,000 nautical miles as various items of equipment had been added. The slipper tanks were flight-tested at Edwards North Base in the summer of 1957, and the modified aircraft were issued to Dets C, B, and A in turn. On 11 October 1957, while it was still dark, Jake Kratt climbed into

Article 351 with its new slipper tanks, and took off from Giebelstadt on Mission 2037.

This would be a non-penetrating flight which used the System 4 payload to pick up ELINT transmissions from the Northern Fleet. After two hours flying over solid cloud, Kratt calculated that he had reached Tromso in northern Norway. He switched the ELINT system on, and turned onto the first leg of a triangular pattern which would take him way out over the Barents Sea and back across northern Norway 3.5 hours later. It was still completely overcast below.

Northern Flights
This was the first time that a U-2 had been sent as far as 70 degrees north. Flying above the Arctic Circle in a single-engine, single-pilot aircraft with only basic navigation aids was a significant challenge. Snow-covered land merged into ice-covered sea, and in any case, polar clouds often prevented a pilot from viewing the earth's surface through the driftsight. The basic standby compass was unreliable so close to magnetic north. The U-2 was also equipped with an MA-1 gyro compass which could be "unslaved" from the magnetic system. This provided an accurate heading signal derived from the gyroscope. But the unavoidable phenomenon of gyroscopic "precession" was a factor, and the rate of precession varied from aircraft to aircraft. The pilot had to perform frequent celestial observations using the sextant; plot the data on his route chart which was divided into grids; and make the appropriate adjustments to his gyro and his heading. But even the best exponents of this "celestial grid navigation" technique could still find themselves 10 or more degrees off course by the end of a long flight, when they eventually flew back over a known point. The technique could take no account of crosswinds which—contrary to perceived wisdom—were not always negligible at the U-2's very high altitude. Moreover, in order to use the precomputed star fixes exactly as provided to him by the mission navigator before take-off, the U-2 pilot had to start the celestial-navigated portion of his flight from a precisely-fixed point and time. If his takeoff was delayed for more than a few minutes, all those calculations were thrown out, and he had to try and replot them inflight.

On the final leg of his planned route across the Barents Sea, Kratt spotted contrails beneath him through the driftsight. He *thought* that he was still in international airspace, but he couldn't be sure. (There was, in any case, a difference of interpretation between the U.S. and the USSR over the definition. The USSR maintained a 12-mile limit, whereas the U.S. was still insisting that sovereign airspace extended only three miles offshore). When he eventually reached the waypoint at Tromso, where the cloud had cleared a little, Kratt determined that he was more than 50 miles south of the planned route. He set course for home, and landed back at Giebelstadt fully nine hours and 53 minutes after takeoff. He had traveled more than 3,900 nautical miles.[2]

The same aircraft was now prepared for a photo mission. The clear weather moved east into the Murmansk area. On 13 October at the same early hour, Article 351 again roared off the Giebelstadt runway, this time in thick fog with Hervey Stockman at the controls. The weather was again overcast on the long flight north, but it began to clear as the U-2 approached the northern tip of Norway. Stockman turned east and flew parallel to the Soviet coast for a while before turning south towards the Kola Fjord. With the three cameras of the A-2 configuration running, Stockman flew across Polyarnyy, Severomorsk, and Murmansk where many submarines, cruisers, and other vessels were moored. Heading further south, Stockman spotted two contrails heading his way, and thought for a moment that Soviet fighters had reached his level—more than 70,000 feet! But the U-2 pilot soon realized that this was an optical illusion; the contrails were far below, as usual, but they were reflecting in the curved canopy above his head, in a "fishbowl" effect. Bill Hall had experienced the same illusion over Volgograd a month earlier. Stockman therefore flew on unpeturbed, and upon reaching Monechegorsk, he turned back to the northwest and left Soviet territory at its short border with the northernmost part of Norway. It took more than four hours to reach home base, where Mission 2040 landed after more than nine hours.[3] When the film was examined at AUTOMAT, the PIs found a MiG-19 captured neatly inflight, directly below the U-2.[4]

A few days later, the Norwegian government sent a protest note about Stockman's flight to Washington. The CIA had not bothered to ask formal permission from Oslo for the aircraft to overfly Norwegian territory. Norway was, after all, a NATO ally in the fight against communism, and had allowed its bases to be used many times for intelligence-gathering flights. The savvy Norwegians had worked out what was going on, from their own SIGINT intercepts of the alert messages which passed between Soviet air defense command posts, as Stockman flew over the Kola peninsula. When he flew back south across Norway, two Norwegian air force interceptors were scrambled from Gardermoen airbase near Oslo, but they were unable to make contact. As a form of compensation, Norwegian intelligence chief Vilhelm Evang demanded copies of the U-2 imagery. The CIA finally agreed, but only on condition that they be made to appear as if the Norwegian air force had taken them! It was another lesson in political reality for Project HQ.[5]

Reshuffle

The two long flights to the north were Det A's swansong. Within a month, the unit was packed up and returned to the U.S. Henceforth, Giebelstadt was used only as a refueling point when U-2s were ferried to and from Detachment B in Turkey (this practice began in February 1958, as a better alternative to disassembly and transport from Edwards by the lumbering C-124s, followed by re-assembly at the Det). Most of Det A's personnel returned to the USAF, including three of the pilots. Colonel "Hack" Mixson, who had replaced Fred McCoy as unit commander, moved to the same position at Det C. There he replaced Colonel Stan Beerli, who moved in turn to DetB where he took over from Colonel Ed Perry. Beerli brought five pilots with him from Japan, and four made the reverse journey. When the reshuffle was over, Project AQUATONE had lost a third of its personnel.

The pilots and other contractors who renewed their CIA contracts were allowed to bring their families to Turkey and Japan for the first time. It helped relieve the monotony, and the relatives who came to Atsugi found an agreeable lifestyle in the well-provisioned naval base, and other U.S. facilities in the Tokyo area. But Incirlik was a less attractive location for accompanied tours. Base facilities were rudimentary, although the married couples were now allowed rent houses off base in Adana. The single men continued to live in the mobile trailers which had been provided for them when the unit first moved in. Beyond the gates lay a barren and sometimes dangerous world, where Kurdish tribesmen roamed, and traveling after dark was a dangerous proposition. Drinking and playing poker therefore remained the favorite pastimes at Det B. Large amounts of alcohol were consumed; substantial amounts of money changed hands.

Fortunately, in addition to the two T-33s assigned for proficiency flying, Det B also had two C-54 transports at its disposal. On some weekends, they were available for shopping and sightseeing trips to Athens, Bierut, Rome, or Wiesbaden. This relieved the monotony. Slowly, too, base facilities at Incirlik improved as more USAF units were assigned there, some permanently and the others on a rotation basis.

More Mishaps in Texas

Three of the aircraft which were made redundant by the closure of Det A were transferred to the USAF. They joined SAC's fleet of U-2As at Laughlin AFB, which was now 18-strong and still growing, as the 4080th wing prepared for the far-flung deployments which had been deemed necessary to conduct the sampling mission. After the two accidents shortly after the U-2s first arrived in June, training at the SAC base continued without a major mishap until late September. The aircraft were still mostly powered by the original -37 engines with their suspect fuel controls and oily air conditioning systems. There were many flameouts, some blamed on autopilot and bleed valve problems, and others on "abrupt maneuvers."[6] SAC's selection criteria for U-2 pilots was rigorous, but the experience level of some was not as great as those selected to join the CIA project. Half of the 4080th U-2 pilots were only First Lieutenants, and the youngest (Tony Bevacqua) was only 24 years old when he was checked out. When the time came to select commanders for the first two sampling deployments, no field grade officers were available, and two Captains were chosen instead (However, they were later replaced by Majors).

On 21 September 1957, Captain Jim Qualls was flying over Nevada on a high-altitude training flight when he flamed out during a 180-degree autopilot turn. He descended to 34,000 feet and managed a relight. Two hours later, Qualls had another flameout and relight over Big Springs, Texas. As the SAC pilot headed for home base, he experienced *another* two flameouts and relights, and was eventually forced to deadstick his U-2 onto a dirt airstrip at a Texas ranch.

The aircraft was recovered to Laughlin and examined. Five days later, U-2 squadron commander Colonel Jack Nole took the same plane up for a test flight. As he climbed through 53,000 feet and deselected the gust control, the nose suddenly pitched down. Nole applied back trim, to no effect. The aircraft entered a dive, and he deployed speed brakes and landing gear in an attempt to regain control. That didn't work, either, and as the engine flamed out the plane went inverted. Nole's mobile control back at Laughlin heard a Mayday call as the pilot decided to bail out. Nole managed to extract himself with some difficulty, but failed to locate and pull the "green apple" which would activate the emergency oxygen system in his seatpack. He was now falling through the stratosphere at 50,000 feet, with his pressure suit beginning to deflate, and no oxygen to breathe. Despite the tremendous windblast, he found his D-ring and elected to pull it. After the parachute opened, he finally located the green apple. Protected by the pressure suit and now happily supplied with oxygen, Nole began a slow, oscillating descent. But the life support system wasn't designed to support such a slow return to the ground—pilots were supposed to free-fall to 14,000 feet, where the parachute would open automatically. By the time he reached 20,000 feet, the oxygen supply was exhausted. Nole disconnected his oxygen hose and began breathing rarified air. He was violently sick, as the parachute continued to swing him back and forth. He eventually reached the ground after a 22-minute descent. It was the highest-ever escape from an aircraft.[7]

The root cause of the accident was determined to be an uncommanded deployment of the flaps, following an electrical short circuit. But Nole's escape also pointed to the need for quick-disconnect fittings for the radio, oxygen, and electrical lines, modification of the parachute, and—once again—an ejection seat.

Crowflight
The SAC wing continued to struggle with training and maintenance problems as it prepared for the High Altitude Sampling Program (HASP), or Operation Crowflight as it became known within the 4080th wing. The HASP requirement, set by the Armed Forces Special Weapons Project (AFSWP), was to determine the quantity of fall-out particles from nuclear weapons tests in the stratosphere; the pattern of their dispersal around the globe; and the extent to which the fall-out was returning to earth through the troposphere.

By now, the U.S. had conducted over 80 nuclear tests in the atmosphere, and 20 Soviet tests had been announced (sometimes by Moscow, other times only by courtesy of the U.S.). There had also been nine British tests. So there was now a good data base, especially since American sampling and diagnostic techniques had been proved by reference to the country's own nuclear test programme. Of course, the patterns of fall-out dispersal from nuclear tests varied considerably according to the weather and the winds. The advent of thermonuclear weapons ("H-bombs") had made the collection process a little more predictable, provided that high-altitude aircraft were available. This was because the fireball from a weapon of megaton yield, being larger and much hotter than that produced by an A-bomb, would inevitably rise into the stratosphere, where winds were negligible and airflow more stable.

During August and September 1957, Lockheed delivered six U-2A-1 models which had been specially modified for Operation Crowflight. They had a small intake door at the tip of the nose, which was opened and closed by a cockpit control. The door led into a duct, which gradually widened to slow the airflow down, until it reached a filter paper placed across the airflow in a ring holder. The newly-developed papers were made from cotton fibres with a gauze backing which were impregnated with an oily substance designed to retain minute particles. Ed Baldwin of the Skunk Works designed an automatic mechanism which could rotate four filters into the airflow in turn, from their stowed position in a circular rack which pivoted at its centre above the duct. By means of an electric actuator operated by the U-2 pilot from the cockpit, the papers could be rotated in and out of the duct in sequence. Since the six modified aircraft lacked the fibreglass panels used for the SIGINT and ADF antennas on other U-2s, they were referred to as "hard nose" aircraft. In late October 1957, the 4080th deployed three of them each to Ramey AFB, Puerto Rico, and Plattsburgh AFB, NY, to start the High Altitude Sampling Program (HASP).

The first phase of flying for the HASP project involved the sampling of a "corridor" along the 70 degree West meridian from roughly 65 degrees North to 10 degrees South, at various altitudes from 45,000 feet to 70,000 feet. It was thought that, over a period of time, most of the earth's stratospheric circulation would pass through this corridor. From each deployment base, the U-2s flew north and south twice each week along the predetermined tracks, usually launching two aircraft at a time to head in opposite directions. A typical flight profile required a level-off at 45,000 feet to collect samples for an hour, then a climb to 55,000 feet and another hour of sampling, then the same from 65,000 feet. Finally, the pilot was required to gain maximum altitude (around 70,000 feet in these profiles) for his final set of samples.

This entailed flying across some quite inhospitable terrain for seven hours or more, over the frozen wastes of northern Canada or the wide open spaces of the western Atlantic Ocean. Three hours before the U-2 was due to take off, therefore, an Air Force search-and-rescue (SAR) C-54 would depart from the Crowflight base and proceed along the flight path. Since the U-2 flew about twice as fast, it would eventually overhaul the lumbering transport, but this procedure gave the maximum

SAR coverage of the route, since both aircraft would reverse course at about the same time. The C-54 carried a para-medical team to assist the pilot if he went down. Unlike the CIA's U-2s, the USAF aircraft were equipped with a rudimentary Collins KWM high-frequency radio, which had to be preset to a single frequency. The pilot was required to make routine calls to the C-54 and the detachment base every so often, otherwise the rescue services sprang into action.

The exposed filter papers were sent to Isotopes Inc in Westwood, NJ, for radio-chemical analysis. Eventually, the results of the analysis were published, and the findings also went to a United Nations committee which had been set up in 1955 to monitor the effects of nuclear radiation in response to growing worldwide concern. As yet more megaton-yield shots were staged, and long-range radioactive fallout spread around the globe, India and Japan led the protests, but there was also concern in Western scientific communities. Two reports from British and American scientists in 1956 attempted to soothe public concern, but even they were forced to admit "the inadequacy of our present knowledge." The presence in long range fallout of the Strontium-90 isotope caused the greatest concern. This fission product remained radioactive for years, and was known to concentrate in the bone structure once it had penetrated the human body. A heavy dose could theoretically cause bone marrow cancer.

The General Takes Command
After deploying the six HASP aircraft, the SAC wing continued to train U-2 pilots and groundcrew, amidst growing concern about how such a unique reconnaissance system should be managed. The peculiar circumstances in which the U-2 had entered the USAF inventory meant that the normal procedures and paperwork so beloved of the military—and especially of SAC—had been side-stepped. Technical manuals and flight handbooks were non-standard; sometimes the groundcrew worked from handwritten notes they had made in the classroom during U-2 ground school at the Ranch. If they wanted a particular part, it came direct from Lockheed rather than through the usual logistics channels. And there was the constant problem of every plane being *different*—it was axiomatic that, for instance, a canopy or a balance arm from one U-bird wouldn't fit another! The "Article" number had to be specified each time a U-2 part was ordered.

In the small, dedicated, and contractor-maintained world of the CIA's Project AQUATONE, these hurdles were easily overcome. In the larger, military-regulated world of SAC's DRAGON LADY program, they were not. The 4080th wing was also trying to keep the troublesome RB-57D operational in a second squadron, plus a fleet of training and support aircraft. A series of incidents focused the attention of higher headquarters on the Laughlin wing. In addition to the U-2 accidents of June and August 1957, two of the wing's T-33 trainers were lost while in the traffic pattern, one at Laughlin and one at Offutt AFB. Two pilots died in the crash at SAC

Headquarters, and two more when a B-57C trainer crashed at Andrews AFB. General McConnell of SAC's 2nd Air Force (to which the 4080th reported) decided to make an inspection at Laughlin. On 5 November, he flew over from Barksdale in his C-54, but as the staff transport entered the traffic pattern at Laughlin, another B-57C making a touch-and-go landed with the wheels up. The runway was blocked, and the fuming general had to divert back to Barksdale! Within a week, McConnell had fired the 4080th wing commander, Colonel "Hub" Zemke, despite his fame and status as a wartime ace. The general decided that the wing needed more standardization and accountability—not less— so he drafted in Brigadier General Austin J. Russell to replace him. Russell had commanded a SAC bomb wing and was a dour disciplinarian in the LeMay tradition. The 4080th's U-2s were shoe-horned into the overall SAC war plan, and in consequence, many flying hours were wasted on filling unnecessary training "blocks."

Russell had only been in charge for ten days when the 4080th suffered another fatal U-2 accident. Captain Benedict Lacombe was returning from a celestial navigation training flight late on 22 November when he crashed on base leg. The accident report concluded that he allowed the aircraft to enter a descending spiral while his attention was diverted. When the warning horn sounded as he exceeded 220 knots, he instinctively applied back pressure on the yoke to reduce speed, but now he exceeded the g-limits and the aircraft broke up. His body was found some way from the wreckage—another U-2 pilot had tried to escape over the side, but failed. By the time that Lacombe was clear of the aircraft, he was too low for the parachute to deploy. Ironically, the new lightweight ejection seat was cleared for service on the U-2 just a few weeks later.[8]

"Hot" Sampling Begins

A different type of sampling deployment now beckoned for the 4080th: Operation TOY SOLDIER. Unlike the HASP, this was a more highly classified effort to gather "hot" samples of Soviet nuclear tests for intelligence purposes. The amount of information about a nuclear weapon that could be deduced from a study of the fallout was quite substantial. Indeed, the very first Soviet nuclear test had only been detected when a USAF RB-29 on a sampling mission between Japan and Alaska had encountered high amounts of radioactivity. This was on 3 September 1949, five days after the event, and a long time before any announcement was made by the USSR. Since then, the esoteric science of "weapons diagnostics" had advanced, so that careful analysis of air-gathered samples could reveal precise details about a nuclear bomb's yield, construction, and composition; the nature of the fusion and/or fission reaction; and whether it had been detonated at ground level, underwater, or in the air.[9]

The Air Force Office of Atomic Testing (AFOAT) used a variety of aircraft to collect samples for weapons diagnostics, but the U-2 was potentially the best platform for high-altitude collection. A new hatch for the U-2 equipment bay was developed and flight tested in early 1957. This contained the P-2 Platform (or "ball sampler package"), which collected gaseous samples by a bleed from the engine compressor, and stored them in six spherical shatterproof bottles. The pilot monitored the pressure build-up in these bottles. After about 50 minutes, upon reaching 3,000 psi, a shut-off valve was automatically activated and the pilot selected the next bottle on his control panel. AFOAT also required particles to be collected as a secondary mission, and the same hatch was also fitted with an airscoop leading to a similar filter paper system as that being used on the "hard-nose" HASP aircraft.

The U-2 "hot sampling" missions had to be flown in the northern latitudes to the east of the Soviet landmass, in order to intercept fall-out from the nuclear test sites at Semipalatinsk and Novaya Zemlya, which topped out in the stratosphere and then began drifting in that direction. The first five ball sampler packages were issued to the CIA for use by Det C, which flew the first missions out of Eielson in June 1957. Subsequently, Det C flew sampling missions out of the home base at Atsugi. On one of these, John Shinn made the first "hot" interception of nuclear debris in a U-2, during a flight going northeast along the Kurile Islands.

The next ten ball samplers were issued to the 4080th wing after it was assigned the primary responsibility for "hot" sampling in late 1957. Operation TOY SOLDIER, SAC's first deployment to Eielson AFB, Alaska, began on 30 January 1958 with three aircraft. The usual area of collection was way to the north, over Point Barrow, where fall-out from Soviet tests on Novaya Zemlya would usually arrive within 24 hours. As in Crowflight, collection was required at various high altitudes, and the missions often lasted eight hours or more. So these flights were long and boring for the pilots. They were enlivened only by the demanding task of polar navigation, and the thought of trying to find a frozen lake to land on, if a flameout forced a descent and landing in the Arctic wilderness!

All through February and into March 1958, the U-2s flew almost daily from Eielson as the pace of Soviet nuclear testing accelerated. Det C also continued to fly sampling missions on a track running north from Japan. The radiation count of the air entering the sampling duct was measured and presented to the pilot on the B/400 radiation exposure meter in the cockpit. There was also a warning light which could be set to come on if the count got too high. The most harmful radiation which was present in stratospheric fall-out was that from gamma rays, which were known to decay within the first 24 hours. It was therefore not usually a factor during the U-2 flights, but if the dosimeter had registered a "hot" count during a sampling sortie, the returning aircraft was washed down with soap and water before the groundcrew were allowed to service it. The samples collected on these flights were sent to AFOAT's laboratory at McClellan AFB, CA, for analysis. There, it was said that scientists could determine a nuclear weapon's characteristics "down to what color it was painted" from the minute samples. Eventually, a laboratory was set up at

Eielson, because the half-life of some radioactive constituents was only a few hours or days, and rapid analysis was mandatory.[10]

More Overflights Requested

The pace of Soviet nuclear and missile testing now alarmed many analysts and politicians in the U.S. Politicans reacted to the launch of Sputnik by orchestrating an anguished national debate about the supposed shortcomings of U.S. missile and space technology. Along the Soviet southern border, the listening stations picked up preparations for another R-7 launch. On 26 October 1957, Dick Bissell went to see Colonel Goodpaster in the White House and requested another U-2 mission over Tyuratam. Goodpaster told him that the President felt it was best to "lie low in the present tense international circumstances."[11]

Eight days later, an R-7 launched the second Sputnik, and U.S. gloom deepened. The CIA still believed that the USSR was two years away from achieving an operational ICBM, but powerful voices in the Congress and the military disagreed. They constructed and broadcast alarming scenarios: what if SAC's nuclear bombers were caught on the ground by a pre-emptive Soviet missile strike? The U.S. didn't even have an early-warning radar network that could track incoming ICBMs. SAC's bomber dispersal and alert programs were stepped up, and development of the Ballistic Missile Early Warning System (BMEWS) was accelerated.

The CIA estimate that the Soviets were still two years away from deploying an ICBM was based partially on knowledge that there were failures, as well as successes, at the Tyuratam launch site. When the listening stations detected preparations for a launch, the CIA's Det B was alerted and a U-2 configured for SIGINT would be launched for a border flight to the east, in the hope of gathering more data if a missile was actually fired. At the same time, a Boeing RC-135 aircraft codenamed Nancy Rae would take off from Eielson AFB and orbit off Kamchatka to monitor the impact area. Often, the countdowns were aborted and the spyplanes returned empty-handed. Occasionally, the missile was launched but failed to achieve the proper trajectory. After the New York Times inadvertently disclosed on 31 January 1958 that the U.S. was eavesdropping on the countdowns, the Soviets limited the extent of their transmissions to the downrange stations. The U-2 couldn't get off the starting block in time, and neither could the Nancy Rae. The U.S. had to enlarge the airfield at Shemya at the westernmost end of the Aleutians, so that the reaction time of the RC-135 could be shortened.[12]

Although the U-2 border flights continued, no further overflights of the USSR were authorized during the winter of 1957-58. However, Project HQ began planning for another high-intensity series like Operation SOFT TOUCH for March or April, when the light and weather over the USSR would be better.[13] But as it transpired, only one overflight was authorized in the spring of 1958. On 2 March, Det C pilot Tom Crull flew Mission 6011 over the Soviet Far East opposite Hokkaido.

He entered denied territory over Sovietskaya Gavan, passing over the naval aviation base there before heading for Komsomolsk and Khabarovsk, where more airbases and aircraft production facilities were situated. Then the flight followed the Trans-Siberian Railroad southwards, close to the Chinese border. Despite using one of the "Dirty Birds," the mission was once again tracked and Soviet interceptors were launched. Shortly after Crull "coasted-out" of denied territory, he had a flameout. Luckily, the MiGs were no longer on his tail.

Four days later, the USSR lodged a strong protest about Crull's flight. President Eisenhower discussed the protest with Secretary of State John Foster Dulles the next day. He had come to the conclusion that these flights just weren't worth the political risk. What if the Soviets misinterpreted an overflight as being an inbound U.S. nuclear bomber?[14] It was all bad news for Project AQUATONE. The Soviet protest proved that they had little difficulty detecting the "Dirty Bird," and the 10,000-foot altitude penalty could have been disastrous on Crull's flight.

Although some work on "stealthing" the U-2 did continue,[15] the Rainbow project was effectively over. The absorbers and the wires worked at some frequencies and aspects, but not at others. Moreover, the foam-based absorber was a maintenance nightmare. In reality, the laws of physics could not be contradicted—the U-2's shape had not been devised with "stealth" in mind. In a design study named Gusto, the Skunk Works was now working on a subsonic, swept wing replacement for the U-2 with no tail, which would have a very low radar cross section. But other studies commissioned by the CIA from SEI, the Skunk Works, and elsewhere, indicated that the best chance of evading radar detection completely would be through more speed and altitude. Therefore, Bissell and his military deputy, Jack Gibbs, asked Lockheed and Convair to submit designs for a supersonic reconnaissance aircraft in early 1958.[16] It would be an ambitious project, with no guarantee of quick results.

For the U-2, meanwhile, altitude was still the best defense, just as Kelly Johnson had always said. As an added precaution against fighter interception, though, the aircraft could be visually camoflaged. In early 1958, a U-2 was painted in dark matt blue for flight trials at North Base. A Lockheed test pilot flying an F-104 in an attempted interception reported that the repainted aircraft was much less conspicuous at high altitude. Some variations of the paint were tried out, but eventually an overall coat of a standard specification known as Sea Blue was chosen.[17] In the summer of 1958, therefore, the CIA's U-2s lost their shiny metal finish, at a weight penalty of 48 pounds. Even Kelly Johnson thought *that* was a price worth paying.

Notes:

[1] Zaloga, "Target America", p 176-181

[2] Don Welzenbach and Nancy Galyean, "Those Daring Young Men and Their Ultra-High Flying Machines," CIA Studies in Intelligence, Fall 87

[3] Welzenbach/Galyean article

[4] Pedlow/Welzenbach, p150

[5] Rolf Tamnes, "The United States and the Cold War in the High North," Dartmouth Publishing, VT, 1991; interview with Vidar Ystad. The Welzenbach/Galyean article, written earlier, states that permission to overfly Norway *had* been granted, but this author prefers the information provided by Tamnes and Ystad. Tamnes is a government historian who consulted the official records in Oslo. Ystad interviewed Evang.

[6] 4080th History, September 1957

[7] 4080th History, February 1958, plus Nole's own account, "I Baled Out Ten Miles Up!" in Reader's Digest, September 1964

[8] 4080th History, February 1958

[9] Moreover, it had even become possible to identify the Krypton 85 isotope which was produced and vented into the atmosphere from Soviet reactors which were producing plutonium. By this means, US intelligence could estimate the other side's rate of nuclear weapons production.

[10] from a history of AFTAC (the successor to AFOAT), quoted in Aviation Week and Space Technology, 3 November 1997, p53, 57

[11] White House Memo for Record, 26 October 1957

[12] Jackson's CIA history of Dulles as DCI, Volume IV, p28-31

[13] notes on special meeting, January 1958, unsigned, White House records

[14] Foreign Relations of the USA, 1958-1960, Volume X, Part 1, USGPO 1993

[15] In a further series of flight tests codenamed BUCKHORN and conducted from Indian Springs AFB, Nevada in June 1958, a different arrangement of the wires was evaluated.

[16] Welzenbach/Pedlow, p263

[17] the specification was actually ANA color 607, "Non-Spectacular Sea Blue". Today, the equivalent specification is FSN 35045

6

New Partners

On a dark evening in mid-November 1957, the British Prime Minister (PM), Foreign Secretary, and Minister of Defence gathered for a briefing at the Air Ministry building in Whitehall. The subject was simple, but extremely sensitive: should the UK respond to the CIA's invitation to become a fully paid-up member of the U-2 project?[1]

When the first U-2 detachment was effectively expelled from the UK in May 1956 by Prime Minister Eden, just a few weeks after it arrived, senior MI6 and RAF intelligence officials were embarrassed. In the early Cold War years, they had painstakingly built close relationships with opposite numbers in the U.S. intelligence community. Both sides had valued the "two-way street" of information-sharing. In the mid-1950s, the Brits still held a lead in some airborne reconnaissance technology, such as side-looking radar and ELINT. But the sheer weight of resources which the U.S. could bring to bear was beginning to tell. After a visit to the U.S. in fall 1956, a British intelligence official warned his superiors that "if the U.S. thought that we were lagging behind, they would not be so forthcoming in exchanging information."[2]

Through the remainder of 1956 and into 1957, the U.S. did continue to provide intelligence from Project AQUATONE to the UK. Art Lundahl of PID was instructed to prepare special briefings, which would be given to senior RAF intelligence officers in Washington. The British officers were allowed to take the materials back to the UK with them.[3] They made sure that their political masters were made fully aware just how important this U-2 intelligence was. "When fully utilized, it will provide us with up-to-date information on the Soviet Air Force which we could not get in any other way, in present circumstances," the Chief of the Air Staff told the Air Minister in late 1956. The Air Minister asked for more details, and passed them on to the PM.[4]

But the British knew they were only getting part of the story. After the first set of U-2 photos of the Middle East were handed over by Lundahl in September 1956, no more coverage of that area was forthcoming.[5] As the "special" briefings to the UK on the results from Project AQUATONE continued in 1957, Bissell sensed the dissatisfaction of the British intelligence establishment. He hatched a scheme to turn this to his advantage. He proposed that the British would join the U-2 project, in return for full sharing of the product.

Bissell reckoned that Whitehall might be more willing to approve overflights of the USSR than the White House, especially now that Sir Anthony Eden had been replaced as Prime Minister (by Harold Macmillan). There was a precedent for Bissell's theory; in 1952 and again in 1954, the Royal Air Force (RAF) "borrowed" RB-45 photo-reconnaissance jets and 100-inch cameras from the USAF, and flew them right across European Russia on missions that the U.S. was reluctant to undertake. In any case, it would be handy to have America's closest ally sharing some of the political risk associated with the U-2 project. Bissell took the scheme to the White House in late October 1957,[6] and exactly a month later, it was briefed to the British PM by an enthusiastic Air Ministry. It involved the induction of a complete British cadre into the CIA's U-2 operation: some pilots and a flight surgeon would join Det B, an operations officer would join Project HQ, and a photo-reconnaissance expert would join Art Lundahl's operation at the PID.

Surprisingly enough, Macmillan agreed.[7] Arrangements were made in the strictest secrecy by the Assistant Chief of the Air Staff (Operations), Air Vice Marshall Ronnie Lees. He established a small project office in the Air Ministry, staffed by a Group Captain and a Wing Commander. Five RAF pilots were carefully selected: they were all in their late twenties, unmarried, and with plenty of experience, including A2 instructor-pilot ratings from the renowned Central Flying School. Upon arriving in the U.S., they were met by Bill Crawford, the same CIA case officer who had escorted all the American pilots through the U-2 induction process. This process was to be no different for the Brits: rigorous medical screening at Wright-Patterson and the Lovelace clinic; then escape and evasion training at Camp Peary, VA. One of the prospective British U-2 pilots backed out at this stage, but the other four made it to Laughlin AFB, where SAC's 4080th wing began checking them out in the U-2 in late April 1958.

Covert Operations

On the other side of the world, the U-2 was getting mixed up in a CIA covert operation for the first time. Since late 1957, the Agency's Deputy Directorate for Plans (DDP) under Frank Wisner had been supporting rebel Indonesian army Colonels who were opposed to the regime of President Sukarno. The White House had approved the covert action, having deemed Sukarno to be too pro-communist.[8] But troops loyal to Sukarno attacked the rebel bases on populous Sumatra. The rebels had another stronghold at Manado, on the northern end of the remote Celebes island chain more than a thousand miles to the east. The DDP decided to step up its aid to the rebels, by providing them with wartime-vintage P-51 fighters and B-26 bombers, which they could use to attack Sukarno's military bases and shipping.

Dick Bissell was happy to lend support to Wisner's covert action. In fact, he reckoned that after two years during which the DPS had amassed tremendous air operations experience of its own, it should actually be running a show like this, rather than the DDP's Air Maritime Division! In mid-March, Det C was told to quickly deploy U-2s from Japan to the Naval Air Station at Cubi Point in the Philippines. Lockheed manager Bob Murphy hurriedly arranged 100-hour inspections on two aircraft, and Det commander Hack Mixson agreed that the ferry flights to Cubi could also serve as the post-inspection functional check flights. Murphy and Mixson flew to the Philippines on a C-47, while 55-gallon drums of the special fuel were loaded and dispatched on a C-124.

Det C began flying missions across the entire Indonesian archipelago on 30 March, looking for the disposition of Sukarno's forces and likely targets for the rebel air force. The DDP wanted the imagery so that it could identify targets for the rebel air force. But it was a huge area to cover; a straight-line roundtrip from Cubi Point to the Indonesian capital at Djakarta was over 3,500 miles. Many of the 30 U-2 missions flown over the next ten weeks were more than nine hours long. Art Lundahl arranged another field deployment to quickly process and interpret the film. Personnel and equipment from PACAF's 67th Reconnaissance Technical Squadron in Japan were sent to Clark AB. After each mission, the U-2 film was flown over to there from Cubi Point in a C-47. It was no mean task to develop and process two spools of B-camera film, each over 5,000 feet long, under field conditions. For instance, the processor only worked at seven feet per minute!

While the U-2s flew intensively, Bissell decided that two of "his" U-2 pilots should join the rebel air force. In late April, two of the project U-2 pilots now based at Edwards AFB with the test unit were dispatched to Manado. Carmine Vito and Jim Cherbonneaux would act as advisors to the Filipinos who had been engaged to fly and maintain the P-51s. But the pair had become pawns in the power-play between Bissell and Wisner: they were not exactly welcomed with open arms by the DDP contingent at Clark AB. As it transpired, their stay was short-lived. The Indonesian air force attacked the base at Manado and, a few days later, a rebel air attack on some of Sukarno's airfields led to the shooting-down of one of the B-26s. Its American pilot was captured, and DCI Dulles ordered an end to the operation.

Overflying Mainland China

The last U-2 flight over Indonesia from Cubi Point took place on 14 June, and Det C began packing up for the return to Japan. However, another long flight over denied territory beckoned for the detachment. Mission 6012 was flown on 19 June out of Naha, Okinawa, when Lyle Rudd became the first U-2 pilot to fly over populated regions of the Chinese mainland. After "coasting-in" over Fuzhou, Rudd flew west across Fujian and Jiangxi provinces before turning north and flying all the way to Beijing. Turning back south, he flew through Shandong and Jiangsu provinces before leaving communist airspace close to Shanghai. Rudd's flight lasted nine hours and 25 minutes. Chinese interceptors shadowed him throughout much of it—obviously their radars were also perfectly capable of detecting a high-flying intruder.

The imagery from Mission 6012 would soon prove useful, as tension between the nationalist Chinese on Taiwan and the mainland communists increased. Two nationalist F-84 fighters were shot down over the Taiwan Straits on 29 July. The U.S. professed its support for the nationalists, and sent them more missiles and fighters. But on 11 August the communists started a bombardment of Matsu and Quemoy, the small offshore islands close to the mainland, which the nationalists still held and had fortified. It seemed for a while that Beijing might be preparing for a full-scale invasion of Taiwan.

But U-2 photography told otherwise. Det C deployed again to Okinawa, from where another mission over the Chinese mainland was flown on 20 August.[9] It revealed a big build-up of fighter aircraft on airfields opposite Taiwan, but no corresponding moves by the communists to build up ground or naval strength in the region.[10] Another mission was flown on 10 September, and for the first time, Beijing issued a protest. By agreement between the State Department, the Pentagon, and the CIA, further flights were suspended. Bissell and DDCI Cabell pressed for another one 12 days later, but State demurred.[11] Eventually, a further U-2 mission over the mainland was flown on 22 October, by which time the crisis was receding. When this penetration was discussed in the White House on 30 October, the President said he was unaware that such flights had resumed, and stressed that permission for them must be secured from him on a mission-by-mission basis.[12]

The last few months had again demonstrated that the U-2 could be used as a "tactical" as well as "strategic" intelligence tool. Not only over Indonesia and China, but also during a renewed series over the Middle East by Det B during the Lebanon Crisis, Lundahl's PID had arranged another field deployment. They provided timely processing and analysis of the imagery. But was "tactical" photo-reconnaissance a role for the CIA? During a discussion with the President on the CIA budget in September 1958, DCI Allen Dulles made a casual suggestion that the Agency's U-2 operation could be transferred, in part or whole, to the USAF. Eisenhower thought some savings could be made, perhaps by passing maintenance responsibility to the Air Force. Bissell, who thought he had won this particular battle twice already, hurried to the White House and scotched the suggestion. He told General Goodpaster that the aircraft should remain "in a small autonomous organization, so as to provide security, direct control, and extremely close supervision."[13]

U-2 safety questioned

According to Kelly Johnson, the maintenance performance at the SAC U-2 wing had been so poor that the Air Force Inspector-General's office had been called in to investigate.[14] The 4080th wing suffered four serious accidents within a month. On the afternoon of 8 July 1958, Squadron Leader Chris Walker was killed when his aircraft went out of control at high altitude and crashed in the Texas panhandle near Amarillo. He was one of the four RAF pilots now in training. The next morning, Captain Alfred Chapin crashed in similar circumstances less than 100 miles away, near Tucumcari, New Mexico. SAC grounded all its U-2s while accident investigations were carried out. Sabotage was suspected, but when the gaseous oxygen systems of other U-2s were checked, excessive moisture was found. Walker's autopsy revealed that he had become hypoxic; it was soon suggested that both accidents could have been the result of the pilots being starved of oxygen, after the supply was restricted by ice formation at the reducer valve. But there were other indications, other mysteries. Apparently, Walker had tried to use the ejection seat, but it hadn't worked. There was evidence of a fire in Chapin's oxygen supply.[15]

On 25 July, a U-2 caught fire on the ramp at Laughlin while the Firewel techrep was checking a new oxygen pressure reducing valve. On 2 August, while investigations continued, the U-2s were cleared to fly again, but only as high as 20,000 feet. Four days later, Lt Paul Haughland was killed on his first U-2 flight when the aircraft stalled on final approach, rolled rapidly to the left, and struck the ground in a near-vertical attitude.[16]

"The airplane has been exonerated," wrote Kelly Johnson in his diary, when the dust finally settled on this unfortunate series of accidents. "We're trying hard to get General Flickenger to take more active steps to go to our dual oxygen proposal," he continued dryly. But Lockheed was assigned some of the blame. SAC insisted that the ejection seats be disabled until Lockheed could assure their reliability. As another precaution, the radio leads to the pilot's helmet were re-routed so that they no longer ran alongside the oxygen tubes: a short-circuit in that wiring could have caused a fire in Chapin's cockpit.

As a result of Haughland's fatal accident, the procedures for training novice U-2 pilots were revised. But the SAC wing complained that the flight manual didn't sufficiently highlight the "unusual stall characteristics" of the airplane, and asked Lockheed to check the aircraft's performance in the landing regime. The 4080th was also still concerned about the autopilots, especially the pitch trim. Eventually, the USAF told the Skunk Works to perform a complete re-evaluation of the U-2's stability and control characteristics. Major Dick Miller, an aeronautical engineer from the Edwards Flight Test Center, was assigned to work with the Lockheed flight test team at North Base, and a new series of flight tests with a specially-instrumented U-2A was agreed. The Lockheed report of this re-evaluation reaffirmed the U-2's basic stability, noting that it met all the important requirements of the basic Military Specification which covered flying qualities (MIL-F-8785).[17]

Twelve U-2s had now been lost in accidents. The CIA had paid for 20 aircraft, and the USAF had taken delivery of another 30, one more than originally planned. Now Kelly Johnson offered the government a sweet deal to produce another five aircraft, which would be partially built from spare parts left over from the earlier contracts. The USAF took up the offer, and the five additional aircraft were delivered between December 1958 and March 1959.

Sampling Stepped Up

The accidents and groundings at Laughlin disrupted plans for the start of a new phase of sampling under Operation Crowflight. Radiochemical analysis of the filter papers from the first series of flights out of Ramey and Plattsburgh had confirmed that fall-out was not mixing across the stratosphere as much as had been thought. It was flowing round and round the earth as expected, but was staying at roughly the latitude at which it had been injected. Even though the HASP sampling corridors were half a world away from some of the test sites, debris from individual shots could nevertheless be identified almost every time the U-2 flew along them, and sometimes within a fortnight of the test shot. Often, the aircraft's radiation exposure meter registered "hot." Another, somewhat disturbing, conclusion was that strontium-90 and other nuclear debris was falling-out of the stratosphere more rapidly than had been anticipated. It was becoming clear that higher concentrations of harmful particles were descending to the ground in certain regions than had been anticipated. Now that the U.S. and Britain were staging megaton-range nuclear tests in the Southern Pacific, a further extension of the sampling corridor into the southern hemisphere was deemed necessary.

After some complicated negotiations, Argentina gave permission for HASP flights to be staged from its territory, although the communist opposition protested. Ezeiza airport near Buenos Aires was chosen as the base, but the three aircraft that departed Laughlin for there in early July only reached as far as Ramey before being recalled after the 8/9 July accidents. They finally flew direct from Laughlin to Ezeiza on 10 September, and Det 4 remained there for the next eleven months. Flights were staged up and down the 64 degrees west meridian, going as far as Trabajares, Brazil (9 degrees south), in one direction, and the Falkland Islands (57 degrees south) in the other. The sampling gear soon picked up debris from the new American and British tests, which were codenamed Operations Hardtack and Grapple, respectively. Meanwhile, the 4080th wing's Det 3 continued at Ramey, from where monthly deployments were made to Plattsburgh to cover the northern latitudes.

Because there was now clearly a weapons diagnostics value to the Crowflight sorties, the six "hard-nose" U-2s at Ezeiza and Ramey were also equipped with the

P-2 platform in the equipment bay. The gas and particle samples obtained from the P-2 went to the AFOAT laboratory at McClellan AFB for analysis. The filter papers exposed in the nose continued to be analyzed in the unclassified laboratory of Isotopes Inc. Also in September, the 4080th wing began a second "hot" sampling deployment codenamed Toy Soldier to Eielson AFB. The pace of Soviet nuclear testing quickened, and the deployment was tasked to mount a flight every day throughout October, using the P-2. Since the three aircraft sent to Alaska had the "conventional" nose with Systems 1 and 3 installed, SAC also took the opportunity to exercise the two SIGINT systems in a few night missions flown closer to Soviet territory than the daytime sampling flights.[18]

The RAF selected a replacement for the unfortunate Walker, who had been scheduled to take command of the British detachment within Det B. He was Squadron Leader Robert Robinson, a test pilot who had been flying the rocket-boosted Canberra. Robinson was therefore already familiar with the high-altitude regime. He arrived at Laughlin in early August. The three other British pilots completed their U-2 training in early October, and together with their flight surgeon and navigator, were soon on their way to Turkey. Every effort was made to conceal the presence of these new arrivals from the Turks and other Americans who worked at the base. The British group occupied their own trailers in the detachment's compound. Meanwhile, the RAF sent Wing Commander Norman Mackie to Project HQ in Washington, where he participated fully in mission planning for all flights, not merely those that would be allocated to the British pilots. Flight Lieutenant Wing Commander Bob Abbott joined the staff of Art Lundahl's AUTOMAT operation, performing U-2 image analysis.

When the British team arrived at Incirlik in fall 1958, a large part of Det B was missing. The detachment commander, five pilots, two aircraft, and support personnel had departed for a secret destination. They were away for over two months. According to the usual strict rules of compartmenting which the Project's CIA security officers laid down, no questions were asked by those that stayed behind.

A Northern Operation

In fact, the missing group was at the Norwegian air force base in Bodo, Norway. Their mission was all the more sensitive since, although the Norwegian government had given permission for them to be there, the main purpose of the deployment had been carefully finessed. That purpose was to overfly the northern USSR looking for offensive missile sites.

Despite its enormous size, the Soviet R-7 missile now being tested from Tyuratam did not have sufficient range to reach the whole of the U.S. when launched from that southerly location on a polar trajectory. In mid-July 1957, therefore, one month before the first successful flight of the R-7, the USSR began constructing its first operational ICBM base in a sparsely-populated forest region south of the Arctic port city of Arkhangelsk. Because of the rail transport imperative for the R-7, the base was located close to the railway line which ran north from Moscow to Arkhangelsk. Great secrecy surrounded the site, and it was codenamed Leningrad-300 to disguise the location.[19] The base was actually known to those constructing it as Angara, but the nearest village was named Plesetsk.[20]

U.S. intelligence had realized that the Soviets needed one or more northern launch bases for their ICBMs. By SIGINT or other means, the CIA apparently detected signs of activity in the Plesetsk area by mid-1958. The analysts in AUTOMAT and OSI were desperately keen to get photos of any potential ICBM launch base, especially during the construction phase. That would give them a known "footprint" by which to compare imagery of the site with other sites where subsequent missile bases might be constructed. Right now, the analysts didn't even know for sure whether the operational launch sites would be fixed or mobile, and hardened or not.

But they *did* know that President Eisenhower was reluctant to approve U-2 overflights. Especially after the latest intelligence fiasco with camera-carrying balloons. Two years after the ill-conceived Project GENETRIX, the President was persuaded by the USAF to sanction a new project designated WS-461L and codenamed MOBY DICK. This time, he was told, a much larger balloon with an improved panoramic type of camera would be launched into the recently-discovered jetstream which ran west-to-east across the USSR at over 100,000 feet. Deputy Secretary of Defense Donald Quarles told Eisenhower that the Soviets would not detect the balloons. These would eventually drop their payloads over Western Europe on command from a timing device. Three of the new balloons were launched from a U.S. aircraft carrier in the Bering Sea on 7 July 1958. Three weeks later, Poland announced that one of the payloads had dropped onto their territory, and a Soviet protest also soon followed. The other payloads were never recovered. Eisenhower was furious.

Despite this, a group of intelligence experts who met with the President at the end of August 1958 recommended that he approve "a northern operation" by the U-2. After checking with Secretary of State Dulles, Eisenhower gave his permission.[21] The flight would enter denied territory from the Barents Sea, and search for Plesetsk and other possible ICBM bases along the railroads of the northern USSR. And now two U-2s were concealed in a hangar at Bodo, the closest-possible "friendly" launch base for such a mission.

Only a half-dozen top government officials in Oslo, and two Norwegian air force officers at Bodo, officially knew that the U-2s were there.[22] The advanced guard arrived at Bodo on a C-130 in late August, led by Stan Beerli, the det commander. They set up base in a remote hangar covered in earth and well hidden on the far side of the airfield. The U-2s were ferried in on 15 September.

To this day, it is not clear whether the Norwegian intelligence chief Colonel Vilhelm Evang knew exactly why the U-2s were at Bodo. If he did know, Evang never told his superiors. They were under the impression that the deployment was related to the new series of Soviet nuclear tests on Novaya Zemlya island, 1,100 miles to the east. (There were a dozen nuclear explosions here in October 1958, followed by a unilateral Soviet moratorium on further tests). Indeed, Det B had brought the P-2 sampling platform with them, as well as the dedicated SIGINT System 4 which was by now in routine use on border surveillance flights from Incirlik. But the Norwegian government's standing policy was to deny the U.S. (or any other country) permission to use its air bases to launch flights which actually penetrated the USSR.[23]

In the small hangar at Bodo, the CIA group waited for a mission plan to be transmitted from project HQ. In Washington, they waited for a break in the weather over the northern USSR. The area of interest was notorious for almost-constant cloud cover. Two long weeks passed, with only a few training flights. A mission was finally alerted...but although it was ambitious, it wasn't a photo overflight. It would be a SIGINT flight all along the Soviet northern coastline, to a landing at Eielson. Jim Barnes was "on the hose" (eg prebreathing) for this mission when it was scrubbed by project HQ. A few days later, Barnes did fly a mission, this time with the sampling platform as the primary sensor. On 9 October, he took off after midnight and flew around the Barents Sea, returning to Bodo after 7 hours 55 minutes. Another two weeks passed before the next alert. This time, it was a SIGINT flight to the same area.

Bob Ericson flew this on 25 October, but it did not go so smoothly. His flight plan was similar to that of Barnes two weeks earlier, with a takeoff before dawn followed by a transit up the Norwegian coast to the Barents Sea, where the U-2's SIGINT systems were switched on. By flying the U-2 as close as 12 miles to the Soviet coast, the hope was that the newly-upgraded recorders would pick up radar activity, as well as communications within the air defense system and the Soviet Northern Fleet. As with previous flights to the area, the navigation task facing the pilot was demanding, especially since it was only daylight during the middle portion of the flight. Ericson reached Novaya Zemlya and identified some of the test installations there through the driftsight as he cruised along the coast. But as he headed back west, he realized that his fuel curve was "out," that is, consumption was exceeding that predicted and plotted by the navigator at Bodo. Ericson decided to abandon that section of the mission which called for him to enter the White Sea inlet. He continued along the coast of the Kola Peninsula instead, but the change of plan meant that his celestial pre-computations were no longer accurate. In fading light, Ericson had to "eyeball" the navigation from hereon.

After flying for eight hours and 25 minutes, the U-2 pilot landed back at Bodo in a crosswind with only 10 gallons remaining. He had made the right decision, although Stan Beerli and mission planners were upset that the White Sea area had gone unmonitored, especially since the postflight read-out of System 4 revealed good SIGINT for the other portions of the flight. Shortly after Ericson landed, another U-2 being ferried into Bodo from Plattsburgh AFB by Tom Birkhead was blown off the runway by the crosswind. Luckily, the aircraft was undamaged and quickly recovered into the remote hangar.[24]

Soviet air defense communications on the Kola Peninsula were routinely monitored by Norway's own ground-based SIGINT operators. They were listening as usual on 25 October. From their intercepts of PVO chatter, they suspected that an unwelcome aircraft had penetrated Soviet airspace, or come very close. The Intelligence Staff in Oslo queried the matter with the U.S., which denied performing any such flight![25]

By the end of October, the long darkness of the northern winter was descending on the region. The chance to photograph the suspected Soviet missile base was gone, and wouldn't reoccur until the spring. The Americans prepared to depart from Norway. To salvage something more from the deployment, one of the U-2 ferry flights back to Turkey was planned as another SIGINT mission. After two alerts and stand-downs, John Shinn finally departed Bodo at 0200 on 6 November and flew east across Finland to the Soviet border. There he turned south and flew along the coast, past Leningrad and the Baltic States, then across Poland and Romania, and out over the Black Sea. A large number of Soviet fighters were launched against this mission, without success. An even larger number of Soviet radars were identified and located by the American ELINT analysts![26]

Nuclear Ransom

In Washington, the missile gap debate was heating up. The National Intelligence Estimate compiled in August 1958 (NIE 11-5-58) suggested that the USSR might achieve an initial operating capability (IOC) with "ten prototype ICBMs during 1959." The intelligence community had detected only six test firings of the R-7 from Tyuratam by this time, and none since April 1958. In fact, there had only been a couple more, plus four Sputnik launches, including one failure.[27] But, a third version of the R-7 had launched the much heavier Sputnik 3 into orbit on 15 May 1958. This satellite weighed over one ton, and therefore clearly indicated that the R-7 could deliver a nuclear warhead (in fact, Korolyev's missile had a nuclear payload of over five tons).[28]

The uncertainties which were implicit in the CIA's analysis of Soviet missile capabilities were seized upon by Senator Stuart Symington, a Democrat who had been Secretary of the Air Force in the Truman administration. Symington was getting different estimates from retired USAF Colonel Thomas Lanphier. He lobbied on behalf of the Air Force Association and also worked for Convair, which was building America's first ICBM, the Atlas. The Colonel claimed to have good contacts in the intelligence community, and claimed that there had been no fewer than 80 test firings, including two with nuclear warheads. He claimed that the Soviets were "two or three years ahead," and would therefore be able to hold the U.S. to nuclear ransom. Symington had taken these concerns to President Eisenhower, and the debate had gone public with articles written by syndicated columnist Joe Alsop, who was also privy to inside information from the intelligence community.[29] Soviet Premier Khrushchev stirred the pot some more, by claiming in public that the production of ICBMs "has been successfully set up."

In fact, Symington and Alsop were getting their inside information from hardliners in the USAF who refused to accept the CIA's estimates, and Lanphier was relying on a former CIA analyst who now worked for Convair. This analyst was discredited within the Agency, but Lanphier had taken him to see General Power at SAC headquarters. Here, the analyst had presented two reports on Soviet ballistic missile deployment, one of which alleged that the USSR had colluded with China and was building railroads into the People's Republic, along which to deploy ICBMs.[30]

On such shaky foundations rested the case of the missile gap proponents. But the cause was enthusiastically adopted by the Democratic Party, which then made substantial gains in the mid-term elections in November 1958. Meanwhile, the intelligence community was obliged to re-assess the August NIE on Soviet ICBM capabilities. The CIA established an Ad-Hoc Panel consisting of outside consultants. The guided missile experts on the U.S. Intelligence Board also reviewed the data. Thanks partly to U-2 overflights and sampling flights, these analysts all knew that the USSR had developed powerful thermonuclear warheads to place on top of the missiles. But they found no reason to change the August 1958 NIE in any significant respect.

In late November, the CIA reported that it still had a "high degree of confidence" that no Soviet missile tests had gone undetected. Why, in that case, did they validate the NIE's prediction that the Soviets would nevertheless have an *operational* ICBM in 1959, albeit only with prototypes and in small numbers? Because (the CIA report continued) of "the Soviets' progress in the whole field of missiles...(the) earth satellite and other (medium-range) ballistic missile launchings."[31] The review was classified at the Secret level, and therefore did not mention the Top Secret intelligence which suggested that the USSR was constructing at least one operational ICBM base, at Plesetsk. The fact that not a single ICBM test launch from Tyuratam had been observed in the previous six months was also omitted, again probably to protect sensitive sources and methods.[32]

Notes:

[1] AIR19/826 file, PRO

[2] CAB159/25 file, JIC meeting with Mr H.S. Young, 18 October 1956

[3] author's interview with Dino Brugioni

[4] AIR19/826 file, PRO

[5] Pedlow/Welzenbach, p114

[6] WH Memo by Goodpaster, 26 October 1957

[7] Brugioni says that the British PM was persuaded by President Eisenhower himself.

[8] Ranelagh, p332

[9] According to Pedlow/Welzenbach, p215, another three missions were flown over China in August 1958. However, these are not listed in other declassified records.

[10] FRUS Volume XIX, item 27

[11] as above, item 79

[12] as above, item 228

[13] WH memo by Goodpaster, 9 September 1958

[14] Johnson log, August 1958. However, none of the maintenance officers of the 4080th wing at the time, recall any such investigation. It's worth noting that the number of accidents and incidents suffered by the SAC wing was a reflection of its far greater flying schedule than the CIA detachments

[15] 4080th history, July 1958

[16] 4080th history, August-September 1958

[17] Lockheed Report SP-117 "Stability and Control Evaluation", dated 1 May 1959; Bob Klinger interview

[18] as above

[19] Zaloga, Target America, p150

[20] sidebar on the history of Plesetsk from an article on Soviet Astronautics, Spaceflight, Vol 38, June 1996, p207

[21] WH memo by Goodpaster, 3 September 1958. Despite the deliberately obscure description - 'a northern operation' - the author believes that this refers to an overflight of the suspected new USSR missile base. This was also the recollection of detachment commander Stan Beerli, interviewed by author.

[22] Rolf Tamnes, "The United States and the Cold War in the High North," Dartmouth Publishing, Brookfield, VT, 1991, p176

[23] Tamnes, p175

[24] Tamnes, p133 and Ericson interview. Tamnes asserts that both the 9 and 25 October U-2 missions out of Bodo were sampling flights, but the author prefers Ericson's account.

[25] Tamnes, p176

[26] Tamnes p176, Bissell memoirs p121, John Shinn interview. Shin n does not recall flying over Eastern Europe during this flight

[27] Zaloga, "Target America", p150

[28] Varfolomyev, p262. Unlike the USSR, the US chose to launch its first satellites on much smaller boosters.

[29] Jackson's CIA history of Dulles as DCI, volume V, section II

[30] as above. The analyst's name has been withheld from the CIA history, as declassified.

[31] CIA "Memorandum to Holders of NIE 11-5-58" dated 25 November 1958

[32] It is not clear how the CIA interpreted the first three Soviet attempts to launch a rocket to the moon between September and December 1958, presuming that these attempts were detected. The Luna program also relied on the R-7 design, to which a third stage was added. The Luna launches on 23 September, 12 October and 4 December 1958 flew for 92, 100 and 245 seconds respectively, before failing. See Asif Siddiqi, "Major Launch Failures in the Early Soviet Space Program", Spaceflight, Vol 37, November 1995, p393

7

Missile Warning

In the winter of 1958-59, thanks to the missile gap controversy, Kelly Johnson nearly managed to sell an astonishing scheme to the Pentagon. He proposed that a fleet of U-2s should be kept on constant airborne patrol around the Soviet borders, to warn of any impending missile attack on the U.S. The "ICBM Detection System" suggested to the USAF by the Skunk Works would have required the building of over 80 more aircraft, in a new version designated U-2B.

The origin of the U-2B proposal can be traced to 13 November 1956. A meeting of the USAF Scientific Advisory Board in Phoenix, AZ, on that date heard Kelly Johnson propose a version of the U-2 equipped with an infrared sensor which could detect the exhaust plumes of jet aircraft many thousands of feet below.[1] A similar IR sensor was also being proposed for missile-attack warning from space by Lockheed's Missile Systems Division in Sunnyvale, CA. This division had received a firm contract to begin development of the WS-117L military satellite a few months earlier. In spring 1957, the Skunk Works received a contract to modify one of the 30 U-2A models being built for the USAF with the IR sensor. It was designated AN/AAS-8 and would be co-developed by Lockheed and Baird-Atomic, the company which had provided the U-2's sextant.

But there were delays with the AN/AAS-8, and it wasn't test-flown until December 1957. The sensor was housed in a large rotating barrel pressurized by dry nitrogen, which protruded from the bottom of the U-2's equipment bay. The barrel contained a mirror which could be tilted in order to focus infrared energy via a second reflector, onto a lead sulfide detector. An infrared oscilloscope was added to the optical path within the U-2's combined sextant/driftsight, so that the data from the IR sensor could be presented to the pilot without installing a new display in the cramped cockpit. The telescope function of the driftsight was made twice as powerful, to achieve x8 magnification and allow the pilot a chance of identifying large intruding aircraft below him.

In this "AIRSearch" configuration, the aircraft was designed only to detect aircraft, as a possible supplement to Airborne Early Warning (AEW) aircraft and the Distant Early Warning (DEW) line of radars. These guarded the airspace over which Soviet bombers would have to fly in any attack on North America. However, fears of a mass Soviet bomber attack were receding, thanks in no small part to the intelligence provided by early U-2 overflights of the USSR. But as the fears of missile attack grew, infrared detection of exhaust plumes remained a subject of great interest. The U-2 with the AN/AAS-8 (56-6722) was delivered in March 1958 to the Special Projects Branch of the USAF Flight Test Center at Edwards.

Project Low Card
This small unit was established specifically to test-fly the U-2 with the new infra-red sensor. Under the leadership of Majors Dick Miller and Robert Carpenter, the Special Projects Branch quickly trained a cadre of four test pilots. On 2 April 1958, Captain Pat Hunerwadel flew the modified aircraft to Ramey AFB for the start of Project Low Card. Flying from Ramey, the sensor's performance could be assessed during the test-launch of U.S. ballistic missiles from Cape Canaveral down the Eastern Test Range. Unfortunately, the aircraft ran off the runway in early June, badly damaging the downwards-protruding sensor. It was airlifted back to Edwards from Ramey for repair, and since the focus of attention was now missiles rather than aircraft, it was decided to rehouse the sensor at the top of the equipment bay. 56-6722 was quickly modified and returned to Ramey in September.

Another lesson was learned from early flight tests of the "IR version," as Kelly Johnson called it. It was asking too much of the pilot, to find and track targets with the sensor, as well as fly the airplane. The Skunk Works therefore modified the equipment bay of the U-2 to carry a second crew member. This "IR observer" would have to fly in even more cramped and claustrophobic conditions than the pilot, though he did also have the benefit of an ejection seat. The first recorded flight of a two-man U-2 was on 7 January 1958, when Bob Schumacher flew Article 377 with flight test engineer Glen Fulkerson in the back. This aircraft was also turned over to the Special Projects Branch. Unfortunately, it was lost in a crash at Edwards on 11 September 1958 which killed Hunerwadel. While trying to land in gusty condi-tions, the wingtip hit hard. The test pilot applied power to go around, not knowing that the skid had been buckled and the aileron balance mechanism damaged. Hunerwadel lost control, and the aircraft spiraled into the ground. Fortunately, no observer was on board. The skid was shortened to prevent it jamming the aileron, if it should be deformed.[2]

In mid-September 1958, Johnson proposed three additional test aircraft, plus another to replace the one which had been lost at Edwards. He also set project engineer Ed Martin and a small team to work on a proposal for an operational ICBM early warning system. In early November, 56-6722 returned from Ramey. It had been in position for 11 of the 12 successful missile launches at the Cape over the preceding half-year. The Skunk Works and Baird Atomic team quickly wrote two reports describing the performance of the AN/AAS-8 sensor in these test flights; missile detection ranges of over 1,000 miles had been demonstrated. In mid-November, Johnson and Walter Baird presented the reports to the Pentagon, and Johnson made a firm proposal for an operational system, with options for 20, 40, or 60 two-place U-2B models. He was asked to fly immediately to Wright Field for further discussions with ARDC. Johnson returned to Burbank with approval for the three extra test airplanes. But the USAF officials had doubts about the communications links and navigational accuracy, especially since the aircraft would be operating within the Arctic Circle. Moreover, they worried about the safety and reliability of the single-engined U-2 in this role.

A second U-2 modified with an observer's position flew in January 1959. This aircraft (56-6954) was one of the five extra U-2s which Lockheed had just built as a supplement to the original 50. The observer was seated facing the relocated oscilloscope, which now also featured a camera, and the control panel which was relocated from the pilot's cockpit.

The U-2B
At the end of January, Johnson submitted a revised proposal for the operational warning system. It was the cheapest and earliest-available solution to the requirement, he wrote. Since the U-2 operated 50% higher than the KC-135 that was also being considered for this warning mission, "this factor greatly enhances its capability in seeing missiles during the launch phase, keeping above weather and refraction effects, and communicating under difficult conditions," Johnson added.[3]

But what a solution it was! To cover all the areas of the USSR from which ICBMs might be launched, three widely separated U-2 patrol areas would be required, with a total of 15 aircraft airborne *at any one time*. Six aircraft each would be launched from Tromso, Norway, and Nome, Alaska, and a further three from Misawa, Japan. They would fly 3,000-nautical mile missions in prescribed loops across the Arctic and the Sea of Japan. The need to maintain line-of-sight UHF communication between each aircraft determined the very large number which must be airborne at any one time (the option of longer-range HF communications was rejected as too unreliable). It took over 1,300 personnel to fly, operate, and maintain the 84 U-2 aircraft which would be required in total. Another 200 people would operate computer centers at the three bases, which would analyze any target data picked up and relayed down the line by the U-2s.[4]

The basic U-2B model provided for a pilot, the observer's position, the IR sensor, an astro-inertial navigation system, and the UHF data link. A conventional tricycle landing gear replaced the U-2A model's unique bicycle configuration; the cost was 884lbs in weight, but "the expected gain in overall operations by lessening landing accidents outweighs this," said the Skunk Works. Even so, the overall empty weight of the U-2B was only 1,500 lbs more than a standard U-2A, at just over 12,000 lbs. Range was cut by nearly 25%, but the U-2B could still reach 70,000 feet. The fuselage length and wingspan were unaltered. To meet the objection to operating a single-engine aircraft so intensively over such hostile terrain, the Skunk Works provided an option to add the small Pratt & Whitney JT-12 turbojet. This provided an emergency get-me-home capability at 35,000 feet, if the standard J57-31 failed and could not be restarted. The JT-12 was housed in the aft fuselage, with the intakes and exhaust covered by aerodynamic fairings to minimize the drag. These fairings had to be ejected before the small turbojet could be lit.[5]

The ICBM Detection System

This fantastic scheme would have cost $200 million to set up, and another $50 million each year to run. But it was seriously considered, and actually won the support of the early warning panel of the President's Science Advisory Committee. This panel met on 11-12 March 1959 to review progress on the Ballistic Missile Early-Warning System (BMEWS) radar and the proposed Missile Defense Alarm System (MIDAS) satellite. The development of MIDAS had been accelerated, but the first test-launch was still many months away, and it wouldn't be in service until mid-1961 at the earliest. Lockheed had promised to have the first U-2 patrol (out of Alaska) operational just 18 months after go-ahead, and all three patrols in service by two years. The panel urged "immediate procurement" of the U-2B system as a complement to BMEWS.[6]

Quite apart from the quick development promised by the Skunk Works, there was an attraction to retaining a man-in-the-loop in any missile warning system—the U-2 observer, in this case. These were early days for IR detection sensors, and their reliability was still suspect. Moreover, Baird-Atomic was promising that the IR sensor on the U-2 could soon be sensitive enough to track missiles even after they had burned out. That offered the prospect of determining the missile's re-entry point over U.S. territory, since just three azimuth and elevation measurements by the observer in one U-2 would be enough to compute a free ballistic trajectory.

But other government advisors were less enthusiastic about the U-2 warning scheme. Moreover, as we have seen, the CIA was not buying the USAF's idea that the Soviets were preparing to launch a pre-emptive strike by long-range nuclear missiles. Another technology suggested by the early warning advisory panel—ionospheric propagation—was already helping the CIA analysts to confirm missile launches. The U-2B proposal was not given an immediate go-ahead, as Kelly Johnson had hoped.

However, the testing of the IR sensor on the U-2 continued throughout 1959 and into the sixties. A third research aircraft joined the Special Projects Branch in late 1959, and spectral analysis of IR radiation emitted by aircraft or missiles was begun, with a lightweight spectrometer which was also developed by Baird-Atomic. As time slipped by, the desperate need to get something—anything—deployed for ICBM detection diminished. BMEWS became operational and MIDAS made progress. The U-2B withered on the vine, and was never built. The U-2B designation was mistakenly used by some people at Edwards to describe the two-seat U-2s used for the continuing IR research.[7]

Promotion for Bissell

On 5 December 1958, Dick Bissell replaced Frank Wisner as the CIA Deputy Director (Plans)—the DDP. Thanks to the U-2 project, Bissell's stature within the Agency had risen rapidly. When it became clear that Wisner was having a mental breakdown, Dulles chose "the golden boy" to take over the CIA Directorate that dealt with spying and covert operations. Richard Helms, who had been in line to replace Wisner, was passed over and remained number two to Bissell in the DDP.[8]

Bissell made sure that the U-2 operation moved with him. His Development Project Staff (DPS, eg Project HQ) previously reported directly to DCI Allen Dulles' office. But on 18 February 1959 the project HQ was formally re-assigned to the DDP. Bissell created a new Development Projects Division (DPD) to manage the U-2, plus all the covert air operations that Wisner had been running. The DPD was headed by Bill Burke, another USAF Colonel who had replaced Jack Gibbs as Bissell's military deputy in DPS in May 1958. Jim Reber remained on the DPD staff, and continued to chair the Ad Hoc Requirements Committee (ARC), which prioritized the targets for the U-2 flights.[9]

DPD stayed put in the old DPS headquarters in the Matomic Building at 1717 H Street NW. But the offices were expanded and divided into various secure compartments which each dealt with a separate project: covert air drops in Tibet, new technical developments, the U-2, and so on. Security was as oppressive as ever. The doors to each compartmented area were kept locked, to deny access to those who were not cleared for that particular project. There was one big briefing room, which all the projects used. Sometimes staff from Project CHALICE—the new codename for U-2 operations—would be told to brief covert personnel from other projects from behind a screen!

Meanwhile, Art Lundahl's Photo Intelligence Division (PID) was also re-organized and expanded. The deluge of imagery from Operation SOFT TOUCH in 1957 led Art Lundahl to conclude that the co-operation between the CIA and the military services within AUTOMAT should be formalized. Instead of the services nominating representatives who would do temporary duty alongside the CIA staffers, but return to their own outfits to write intelligence reports, Lundahl suggested permanent assignments. The Army and Navy agreed, but (as usual) the USAF guarded its autonomy. The Air Staff insisted on retaining an independent photo-interpretation group in the Pentagon (in addition to the large 544th RTS operation at SAC headquarters). Despite this, Lundahl's plan was implemented, and the PID became the PIC (Photo Interpretation Center). Henceforth, Lundahl reported directly to the CIA Deputy Director for Intelligence (DDI), Robert Amory. Despite its new status, however, AUTOMAT continued to occupy the obscure Stueart Building in Washington's run-down northeast suburbs.[10]

The elevation of Bissell to become DDP represented a significant victory for technical intelligence over human intelligence within the CIA. In addition to the U-2 and the covert air operations, the expanded DPD was also now in charge of two hugely ambitious aerospace technical developments. The first of these was the continued search for an air-breathing successor to the U-2. The CIA had now rejected some of the more unlikely proposals, such as hydrogen-fueled airplanes, a huge aerostat which could fly above 100,000 feet, and a ramjet-powered vehicle which would be towed to 60,000 feet behind a U-2 before the ramjet was lit. A new panel of scientists, once again chaired by Din Land, was helping Bissell evaluate two competing proposals for Mach 3 air-breathing vehicles from Lockheed and Convair.

The second top-secret development which DPD now managed was CORONA—the project to field a photo-reconnaissance satellite as rapidly as possible. In February 1958, President Eisenhower had approved a plan to "hive off" part of the USAF's large and multipurpose WS-117L satellite program, and establish it as a CIA-led program. CORONA would use a panoramic camera which was scaled up from the one used in the USAF's unsuccessful MOBY DICK reconnaissance balloon. The film, which was exposed in space, would be returned to earth unprocessed, in a reentry capsule. The USAF's preferred alternative of scanning the film electronically in space, and sending the results to earth via a datalink, was more technically ambitious. Therefore it was less likely to produce early results.

Bissell and the DPD had responsibility for getting the "interim" CORONA satellite system into operation as quickly as possible, to provide an alternative window on Soviet strategic capabilities. This move reunited Dick Bissell and Ozzie Ritland, who was by now a Brigadier-General at Ballistic Missile Division in Los Angeles. Bissell and Ritland became the senior managers for CORONA within the CIA and USAF, respectively—just like their partnership in the early U-2 days. In another echo of the U-2 program, a cover story was devised to hide the true purpose of CORONA. The planned launches of the satellite on a Thor booster mated to an Agena upper stage were said to be for research purposes, in a series named Discoverer. The first Discoverer launch took place on 28 February 1959, only 12 months after the program go-ahead. It failed.[11]

CONGO MAIDEN

Bissell's empire had grown by leaps and bounds, but he hadn't been able to prevent one particularly sensitive U-2 mission from being claimed by the USAF. The blue-suiters emerged from the renewed discussions of September 1958 over who-does-what, with the responsibility for peripheral photo-reconnaissance. That is, flights which gathered imagery intelligence but were not intended to overfly denied territory. On 20 March 1959, SAC deployed three U-2As of the 4080th wing to Eielson AFB to conduct photo missions codenamed CONGO MAIDEN along the eastern and northern coastline of Siberia. The detachment was led by Colonel Hayden "Buzz" Curry, and used three of the five most-recently built U-2s from the supplementary batch.

The main purpose of this operation was to determine the status of Soviet airfields in this frozen region. Many of these had been built along the coast before WWII, and six of them were known to have been upgraded in recent years, despite the difficulties of constructing paved runways and buildings in the permafrost. The upgrading of the Arctic airfields, especially those in the Chukotsky Peninsula, suggested that Soviet nuclear bombers might use them as staging posts, on their way to attack prime targets in North America. SAC also wanted to find out more about Soviet air defenses in this region, in case a pre-emptive U.S. strike on the Arctic airfields was ordered. The mission planners at Offutt set to work on the flight routes, which would be flown 12-15 nautical miles off the coast, using the B-camera in Modes 3 and 4, which provided coverage from the vertical out to the right or left horizon, respectively. The U.S. still formally insisted on its right to fly as close as three miles to the Soviet (and any other) coast. But international sentiment now favored 12 miles. The mission planners at SAC headquarters deliberately did not press for three miles, since they didn't want the politicians in Washington to sit up and take notice of these flights. In addition to the B-camera, the aircraft would also carry the small ELINT (System One) and COMINT (System Three) receivers in the nose.[12]

The window of opportunity for mounting such photo flights was limited, not only by the darkness of the northern winter, but by frequent fog and cloud in summer. But in the second half of March 1959, the weather was unusually fine in the area of interest, with only some high cirrus cloud. The SAC detachment quickly prepared for the missions, and on 23 March Major James "Snake" Bedford took off towards the Bering Strait on the first one. Over the next five days, another six missions were staged, heading southwest down the Chukotsky and Kamchatka peninsulas, and northwest into the Chukchi and East Siberian Seas. The navigation was challenging: through the driftsight, the pilots had great difficulty distinguishing frozen snow-covered land from frozen, snow-covered sea. There was precious little vegetation on the land. Once again, celestial navigation was essential. But it would not have been surprising if the 15-mile limit was breached on a few occasions, given the well-known difficulties of navigating in the Arctic North.

The longest mission of this peripheral photo series was flown by "Buzz" Curry himself on 27 March, a nine-hour flight which went to the Kamchatka peninsula and was a 100% success. The other pilots who made these long and lonely flights were Captains Rudy Anderson, Bobbie Gardiner, and "Snake" Bedford. The only snags were with the tracker cameras, which failed on two flights. However, the PVO had radar coverage even in these frozen wastelands, and MiGs were stationed at some of the bigger airfields, such as Anadyr, Mys Shmidta, and Provideniya. It seems likely that at least some of the CONGO MAIDEN flights were detected. After completing the task, the SAC detachment remained at Eielson and conducted a new series of sampling flights from there until mid-May.

More Small Wars

The record for the longest U-2 flight in 1959 went to the CIA's Detachment C, thanks to its involvement in the crisis over Tibet. A rumour that the communists intended to abduct the Dalai Lama, spiritual and temporal leader of the Tibetan people, led to a full-scale rebellion in that remote land in March 1959. The CIA had been lending active support to the rebels since fall 1957. But the Red Army was firmly entrenched, and the Dalai Lama and many of his supporters were forced to flee across the border into India, when the March rebellion failed.[13]

In early May 1959 the Eisenhower administration authorized a major expansion of U.S. support, including the training of 700 Tibetans at a special operations camp in the mountains of Colorado. After training, the nationalists would be airdropped into Tibet, with the aim of disrupting Beijing's supply lines to the remote region.

More imagery of Tibet was needed to identify suitable drop zones and targets for sabotage. The U-2 was ideal, since mission planners in project HQ had other targets near the Sino-Soviet border which Jim Reber and ARC had assigned high priority. A portion of Detachment C was redeployed to NAS Cubi Point, and flew two long missions north from there, all the way across China and Mongolia. Tom Crull flew the first on 13 May 1959, a nine-hour ten-minute trip which reached as far north as Mongolia before turning south to fly across Tibet and land at Kermitola, a military airbase near Dhaka in East Pakistan.

Two days later, Lyle Rudd flew even further north. He passed over Ulaan Baatar, Mongolia, and then followed the railroad north to Ulan Ude, crossing into the USSR and flying as far as Lake Baikal. Finally, Rudd turned back to the south for a long tedious leg across the barren Chinese steppes and deserts. Rudd eventually reached Lhasa, and had a superb view through the driftsight of the Potala Palace in the late afternoon sunshine. Like Crull before him, Rudd recovered into Dhaka airfield in East Pakistan as planned. His low-fuel warning light came on even before he found the airfield. Rudd landed after nine hours 40 minutes in the air, which was probably the longest operational U-2 mission to date. He had covered over 4,200 nautical miles.[14]

The film from these two missions was processed and analyzed by interpreters from PIC who deployed to Okinawa. The USAF had agreed to provide C-130A transports based there for the airdrops. These began in July 1959, staging out of a base which DPD had secretly established at Takhli in Thailand.[15]

The following month, Det C deployed U-2s to the new Thai airbase. Three flights over China and Tibet were mounted in early September.

From Takhli, Det C also flew over Laos and North Vietnam. The Geneva Agreement which supposedly guaranteed the neutrality of Laos was being broken daily by North Vietnam, but also by the U.S., which sent "military advisors" to this small, landlocked kingdom. The "advisors" were apparently controlled by DDP. On 7 September 1959, the pro-western government in Vientiane asked the United Nations to intervene because rebels trained by the North Vietnamese had invaded frontier areas. Det C made three U-2 flights across the area between 30 August and 8 September. Photo-interpreters derived the North Vietnamese order of battle from the imagery, but were unable to determine the extent of infiltration into Laos, due to the thick jungle canopy covering the border areas.[16]

Was it appropriate to use the nation's primary strategic intelligence-gathering tool in these covert actions? Din Land did not think so. He was still a key advisor to the CIA, chairing the panel that was advising on the satellite and U-2 replacement projects. Eventually, he complained strongly to Dick Bissell about the continuing use of the U-2 in the CIA's clandestine actions around the world.[17]

Even so, everyone in DPD knew that the Soviet Union was still the primary target. Where were the missiles? Could the U-2 find them? Would the President let it fly? Most important of all, could the Soviets shoot it down?

Notes:

[1] Johnson diary, 13 November 1956

[2] interviews with Bill Frazier, Dick Miller

[3] Clarence Johnson, "ICBM Detection System", Lockheed Report SP-112, 26 January 1959, Summary

[4] as above, Sections 2-4

[5] as above

[6] 13 March 1959 Report of the Early Warning Panel, President's Science Advisory Committee, in White House records, Office of the Special Assistant for Science and Technology, DDEL.

[7] Various authors have attributed the U-2B designation to other sub-variants of the U-2, but they are all mistaken. Even this author was guilty, in "Dragon Lady", of speculating that the U-2B was the bomber version proposed by Kelly Johnson back in 1956. Not so! Incidentally, the two-seat U-2s described here, which were used for the continuing IR research project by the Special Projects Branch after the ICBM warning system proposal failed, were eventually designated U-2D models in 1960 or 1961.

[8] Ranelagh, p328-9

[9] CIA Notice No 1-120-2, "Organization and Functions, Office of the DDP", 18 February 1959, declassified 1990

[10] CIA Notice No 1-130-5, "Office of the DDI - PIC", 19 August 1958, declassified 1990; interviews with Bill Crimmins, Art Andraitis

[11] Dwayne Day, John Logsdon, Brian Latell, "Eye in the Sky - The Story of the CORONA Spy Satellites", Smithsonian Institution Press, Washington, 1998, p5-6

[12] interview with Orville Clancey

[13] Christopher Robbins, "Air America", Avon Books, New York, 1985, p74-76

[14] author's interviews with Lyal Rudd, Tom Crull. For some reason, the CIA does not list Rudd's mission as an overflight of the Soviet Union. Maybe it was not explicitly briefed as such to President Eisenhower, in case he refused permission. The Soviet targets could have been the Angarsk nuclear weapons plant and other military-industrial facilities around nearby Irkutsk. Or, this portion of the mission may have been designed to prove or disprove the theory which had been recently advanced by some missile gap proponents, that the USSR had conspired with China to hide ICBM sites along the railroads which crossed into the People's Republic.

[15] William Leary, "Secret Mission to Tibet" in Air & Space magazine, December 1997/January 1998

[16] Brugioni book, p39

[17] author's interview, CIA source

8

New Model

The Soviet fighter threat to Project CHALICE was growing. PVO pilots were practicing the "zoom-climb" technique. After climbing to near maximum altitude, they accelerated in the straight and level, or even a shallow dive to maximum speed, before pitching up sharply. The interceptor entered a ballistic curve, gaining a few thousand extra feet. The maneuver was risky, since the fighters' control surfaces had little effect at the top of the curve. But in theory, the interceptor pilot might be able to bring weapons to bear on a high-altitude target, if the zoom-climb had been entered at exactly the right spot, relative to the target.

Meanwhile, Soviet fighter designers were working on new interceptors which could unambiguously reach the U-2. The MiG bureau had developed a delta fighter designated Ye-50, capable of climbing to 65,000 feet in 9.4 minutes thanks to a booster rocket. It was short on range, though, and plans for a batch of 20 production aircraft, with a large centreline droptank for extra fuel, were dropped. MiG also added a ventral rocket pack to the MiG-19, and designed a version that was rocket-launched from a sled. Also, by late 1958 MiG's new lightweight delta design was evolving from the experimental Ye-5 and Ye-6 configurations, into early series production, as the MiG-21F.

These MiGs were not mature weapons systems, however, and their Tumansky powerplants were not fully developed. But a new version of the now-familiar MiG-19 was now entering mass production for the PVO interceptor regiments. Codenamed *Farmer-B* by the West, the MiG-19PM was credited with a ceiling of 60,000 feet, and it also carried an early Soviet airborne search radar, the RP-5 Izumrud (Emerald). This rudimentary system had two antennas, one scanning 40 degrees in azimuth to identify the target, and another which issued commands to guide a beam-riding air-air missile (designated K-5 or RS-2U.)[1] The Izumrud had been identified by Western ELINT systems, and was correctly assessed as having a search range of only 10 nautical miles, and a maximum track range of five miles. Even so, the *Farmer-B* might now be a threat to the U-2, if the interceptor pilot could get a radar lock-on for the missiles and fire them upwards. Of course, if the U-2 was forced to descend for a relight after flaming-out, the threat would be that much greater.

The new missiles and radar were also fitted to a new interceptor designed by Pavel Sukhoi, which used the more reliable and powerful Lyulka AL-7 axial turbo-jet. The Su-9 evolved from Sukhoi's T-3 experimental delta, which had been flown and revealed in May 1956. The T-3 was capable of Mach 2 and could reach 65,000 feet. But Western intelligence knew very little about the Su-9: although it entered production in 1959, the codename *Fishpot* was not even allocated until 1961.

New SIGINT systems

It was obvious, though, that the U-2 was becoming increasingly vulnerable. The more so because each time a new piece of equipment was added, the aircraft got heavier and its maximum altitude declined. The new equipment included an improved SIGINT system which used new antennas designed by HRB Singer under contract to the CIA. System 6 was introduced in 1958 to replace the rudimentary System 1 ELINT package on the Agency's U-2s. It provided continuous coverage of a wider frequency range without the need to preselect certain bands before take-off. Small spiral and parabolic antennas were installed behind flush-mounted radomes on the equipment bay hatches, which could nevertheless still carry the primary A or B camera payloads. A larger scimitar antenna was added beneath the rear fuselage, enclosed in a rectangular radome. This antenna was shared by System 3, the COMINT system, which had previously used a nose-mounted antenna. Responsibility for the overall design and maintenance of the U-2's SIGINT systems now rested with STL, a subsidiary of Ramo-Wooldridge that survived the 1958 breakup of that company.

By now, moreover, U-2 mission planners fully realized that the maximum altitude of the U-2 could also be reduced by unexpected variations in the outside air temperature at high altitude. The problem was particularly acute in the northern latitudes. This was because the tropopause was lower towards the pole than it was at the equator—sometimes as low as 25,000 feet. Above the tropopause, air temperature rose steadily again, so that by 60,000 feet it could be 5-10 degrees Centigrade warmer than the standard (which was -55 degrees). Under these conditions, the U-2 climbed slower, and even failed to climb at all if the temperature was more than 10 degrees warmer than the standard.

Considering that the northern regions of the USSR were prime intelligence targets, this was bad news. Kelly Johnson concluded that a more powerful engine was needed, to boost the U-2 back above 70,000 feet. Fortunately, the Pratt & Whitney J75 engine was the solution. This was a scaled-up version of the J57, which had entered production in 1957 to power the USAF's latest fighters, the F-105 and F-106. Although mass airflow was increased by one-third, the J75's dimensions were only 10 percent greater than the J57.

A quick study confirmed that the J75 would just fit in the existing U-2, if the width of some main fuselage rings was reduced. The CIA was offered nine J75P-2 engines which had originally been intended for the Martin XP6M Seamaster, a jet flying boat which the Navy was about to cancel. They were converted for high altitude flight by modifying the hot section and fuel controls. In the new J75P-13 configuration they offered 17,300 lbs thrust, which was a huge increase compared to the U-2's existing J57-31A. A few other modifications to the airframe were necessary to accommodate the heavier, more powerful J75. The intake duct had to be widened, and the main landing gear bulkhead was strengthened to cope with the higher gross weight. The modified aircraft would be 1,450 lbs heavier than the U-2A, with the engine itself accounting for 1,100 lbs.[2]

Eisenhower's Doubts

Development of the new U-2 model took place in the spring of 1959. In Washington, though, the prospects for further U-2 overflights of the USSR were very uncertain. Despite persistent prompting by senior intelligence officials, President Eisenhower was as reluctant as ever to approve them. Just before Christmas 1958, Ike told his Board of Consultants on Foreign Intelligence Activities that he doubted whether "the intelligence which we receive from [the U-2 flights] is worth the exacerbation of international tension which results." Moreover, he believed that the previous flights had already provided plenty enough strategic targets for SAC's bombers and missiles.[3]

After a meeting of the National Security Council on 12 February, Defense Secretary Neil McElroy, his assistant Donald Quarles, and Joint Chiefs of Staff chairman General Twining all stayed behind. Just a few days earlier, Premier Khruschev claimed that series production of an ICBM was underway in the USSR. McElroy pressed for more flights, reminding Eisenhower that not a single launch base for operational Soviet ICBMs had been found. Twining backed him up, claiming that the risk of a U-2 being shot down was remote. Top officials in the Pentagon were still disputing the conclusions of the high-level review of the missile intelligence estimating process which had been undertaken the previous fall. According to them, the reason that the Soviets had not test-flown any more ICBMs from Tyuratam for nearly a year, was that they were satisfied with the big missile's performance! Ergo, they were now deploying it.

Ike's response was to doubt whether the Soviets could actually build operational launch sites so quickly. "We generally overestimate the capability of the USSR to outperform us," he noted, reminding the meeting of how the threat of a "bomber gap" had been so exaggerated a couple of years earlier. He still thought that the U-2 flights were provocative: "nothing would make me request the authority to declare war more quickly, than violation of *our* airspace by Soviet aircraft," he declared. Ike wanted to wait for the successor airplane to the U-2, which the CIA was striving to develop, with its promise of undetectability. Better still, wait for the reconnaissance satellite, which did not violate Soviet airspace at all.

Ike left open the possibility of approving one or two flights, though. At this point, Colonel Goodpaster reminded him that he had already approved a mission over the northern USSR to find and photograph suspected ICBM bases. It had not been flown because of the unfavorable sun angle and weather. (This was the mission which Det B had attempted to fly out of Bodo during the previous October). This mission was the number one priority, Goodpaster continued, and conditions to fly it should be favorable again from March. Twining emphasized how important this flight was, and Eisenhower reluctantly agreed. "At least we'll then find out whether the Soviets have an adequate surface-to-air missile," he noted dryly. Twining quickly reminded Ike that the Soviets had not yet fired one of these missiles at any U.S. reconnaissance aircraft. Quarles chipped in, assuring the President that only a few overflights of the USSR would be planned, and each one would be brought to him for clearance. The meeting broke up, with Eisenhower reminding those present that the USSR was ratcheting up the tension over the status of Berlin.[4]

Mission planners once again reviewed the options for "the northern flight," and other high-priority targets. But their efforts were frustrated within days, when the President told Colonel Goodpaster that he had decided to disapprove any additional U-2 flights because of the crisis over Berlin.[5] Still, the U.S. intelligence community, led by Bissell, pressed for a resumption. Their cause was strengthened in March, when the USSR finally resumed ICBM test-flights from Tyuratam. (In the meantime, however, there had been three unsuccessful Soviet attempts to launch a probe to the moon from there, followed by another propaganda coup when the Luna rocket did finally achieve success on 2 January 1959. The Luna was based on the R-7 ICBM, with the addition of a new third-stage).[6]

More U-2 flight plans were laid out, reworked, and refined within the secure confines of 1717 H Street. On 3 April, DCI Allen Dulles himself discussed them in the White House. As well as the much-prized "northern flight," there was a proposal for another flight over Tyuratam. Both flights would cover a number of other high-priority targets.[7] President Eisenhower was reluctant to approve them, but said he would consult with Acting Secretary of State Christian Herter. Three days later, Ike did give a go-ahead for the flights, after his trusted advisor Colonel Goodpaster said he favored them taking place. But within 24 hours, Eisenhower had changed his mind. On 7 April, the President called in Dick Bissell and Secretary of Defense McElroy and told them that the flights weren't worth the political risk. Ike agreed that the missile intelligence was needed, but expressed his concern over "the terrible propaganda impact if a reconnaissance plane were to fail." He thought that the Soviets were ready for a Summit Meeting, which he didn't want to jeopardize.

A frustrated Bissell returned to the Matomic Building. He never told the staff there exactly what transpired in any of the White House meetings, but they could guess. Each time a U-2 flight was proposed, they would make up detailed briefing boards with maps of the route, and send it to the Executive Office. The maps were accompanied by a cogent, one-page summary justifying the flight. This was crafted by Jim Reber, with help from CIA analysts in specific fields, including Herb Bowers (strategic bombers), Sidney Graybeal (missiles), and Henry Lowenhaupt (nuclear weapons). Usually, DPD's military boss, Colonel Bill Burke, would take the briefing to the White House Staff Secretary, Colonel Andrew Goodpaster. He would subsequently take it in to the President. But "Speedy Gonzales"—as the DPD staffers called Eisenhower—was obviously not swayed by the briefings. There had not been a single deep penetration overflight of the USSR for 18 months.

U-2C First Flight
On 13 May 1959 at Edwards North Base, Ray Goudey made the first flight of a U-2 re-engined with the J75. Article 342 was the prototype, and the conversion received the new designation U-2C. An intensive series of test flights followed, in which the advantages and disadvantages of the modification were quickly explored. On the plus side, the maximum altitude was raised to more than 75,000 feet, and the new engine had a much greater compressor stall margin. There were no more bleed valves to worry about, either. But range was sacrificed for altitude: a reduction from 4,000 to 3,300 nautical miles when a maximum power cruise profile was flown. This was the favored profile, however, since the U-2C entered the cruise-climb at the higher altitude of 67,000 feet, and nearly two-thirds of the mission was flown above 70,000 feet.[8]

On the minus side, the re-engined aircraft's center of gravity (cg) limits were narrower. On an early test flight, Lockheed pilot Bob Schumacher reported that the cg was so far forward that the aircraft might become uncontrollable if the autopilot disconnected before the slipper tanks had emptied. The horizontal tail was modified with a more rounded leading edge and an increased camber, to balance the aerodynamic load. Even so, ballast had to be added at the tail for most operational configurations. However, Lockheed did take the opportunity to incorporate three improvements for controllability of the C-model, based on earlier experience with the A-models. The elevator trim tab operating speed was doubled, and the down elevator travel was increased from 11.5 to 20 degrees to offset the nose-up pitching tendency during rapid application of power. The speed at which the flaps and ailerons moved to and from gust control position was halved, to decrease the rate of change of elevator stick force.

The Israeli Bomb

In Turkey, Detachment B's only operational flying was occasional SIGINT flights along the Soviet border, and the photo missions over the Middle East and the Mediterranean at roughly monthly intervals. The Lebanon Crisis of mid-1958 and the overthrow of Iraq's pro-Western monarchy taught U.S. policy-makers to keep an eye on the increasingly-volatile region.

The photo-interpreters at AUTOMAT began noticing a lot of activity at an Israeli air force bombing range south of Beersheba. Near the small desert town of Dimona, a large barren area was fenced off and a new access road was built. From successive U-2 flights, the PIs noted the arrival of construction machinery, and then the digging of heavy foundations, which were reinforced with concrete. They began to suspect that Israel was secretly building a nuclear reactor. After the latest imagery showed that concrete footings had been poured—surely for a reactor dome— Art Lundahl took the evidence to the White House. There was no immediate reaction: the U.S. was quietly supporting Israel against its pan-Arab, pro-Soviet neighbors.

Throughout 1959, Lundahl continued to report the activity at Dimona to the White House and to the Atomic Energy Commissioner, Lewis Strauss. Construction of a second underground site began nearby, and the CIA analysts concluded that it would be a chemical reprocessing plant to make weapons-grade plutonium. They noticed a striking similarity between Dimona and the French nuclear reactor and processing facility at Marcoule...

The Israelis began reacting to the U-2 overflights. They launched interceptors, but without success. The CIA briefed U.S. military attaches in Tel Aviv, who began snooping around the perimeter of the desert site. By the end of 1959, there was no doubt that Israel was developing a nuclear weapon. But the White House seemed determined to look the other way.[9]

More Soviet Border Flights

With the resumption of missile tests from Tyuratam, the U-2's ability to collect electronic intelligence along the Soviet southern border assumed new importance. Whenever there were indications from the ground stations monitoring Soviet communications that a test was planned, an aircraft was made ready at Incirlik. The pilot who was "in the barrel" (that is, rostered to fly the next operational mission[10]) was told to standby. If the indications continued, he was summoned to the operations building, and would sleep there if necessary. The next stage of alert required him to be suited-up, and placed in the cockpit where the prebreathing process was completed. It could be hours before a launch order finally came through. More often than not, though, the mission was canceled.

These "strip alerts" were also endured by another set of crews on the Incirlik flight line. They were from SAC's 55th Reconnaissance Wing, flying a version of the B-47 bomber which was specially configured for the same TELINT (Telemetry Intelligence) mission. Two of these aircraft, codenamed Tell-Two and designated EB-47TT, were also alerted when the first indications that there might be a missile launch were received. The primary aircraft would depart to the east and set up an orbit over northeastern Iran, providing another means to intercept the countdown. If the countdown was delayed for some reason, the secondary aircraft was available to relieve it.[11] But the Tell-Two could not fly high enough to intercept any telemetry from the vital first-stage burn of Soviet missiles launched from Tyuratam. Nor could it venture beyond Iran's eastern border with Afghanistan. That was where the U-2 came in. If the U-2 was actually launched, it would fly higher and further than the B-47.

Most U-2 missions of this type went as far east as the Afghan-Pakistan border and lasted up to nine hours. Many were unsuccessful, as the Soviets aborted the countdown. On one frustrating occasion, John Shinn loitered as long as he could at the turnaround point before turning for home. Three hours later, the countdown resumed, and the missile was launched just after the U-2 landed back at Incirlik. Finally, though, the elaborate preparation and co-ordination paid off. On 9 June 1959 the first-stage telemetry of the R-7 was intercepted for the first time during U-2 mission 4120.[12]

Operation TOUCHDOWN

The renewed activity at Tyuratam may have persuaded President Eisenhower to finally heed the urgings of his advisors, and approve a new overflight of the launch site. Det B was alerted while Allen Dulles and Bissell brought a plan named Operation TOUCHDOWN to the White House on 7 July. It called for a takeoff from Pakistan, in another attempt to penetrate Soviet territory without detection. The flight would cover the Saryshagan test range and the Semipalatinsk nuclear test site again, followed by the nearby Dolon airfield, where long-range *Bear* bombers were based. Then it would head northwest towards the Urals, where many military-industrial targets were on ARC's list. The flight would then turn back from Sverdlovsk to the south, overfly Tyuratam, and depart Soviet territory at the border with Iran and Afghanistan.

For the first time, a landing would be made in Iran. The Shah was being supported by the CIA, and had no objections. Det B commander Stan Beerli and executive officer John Parangosky had flown over Iran in a C-47, looking for suitable landing sites. They selected an airstrip near Zahedan. It had an old wartime runway, a radio beacon, and little else! It was suitably remote, but there was some concern that the place was not secure from mountain bandits who operated freely in the area.

Eisenhower liked the idea of entering and leaving Soviet airspace at widely separated points in the south. But he withheld final approval until Secretary of

State Herter could be consulted. At a second meeting in the Oval Office to discuss the flight next day, Herter strongly supported Dulles and Bissell. Ike bowed to their combined pressure, but noted that "a complete disavowal" would have to be made, if the Soviets protested the flight.[13]

The "QuickMove" procedure swung into action. Colonel Stan Beerli, the Det B commander, had devised this fast and stealthy means of deployment using a couple of C-130s. All the items necessary to support a mission—commo, PSD, and maintenance gear—was packed onto three trailers which could be easily pulled on and off the C-130 by a jeep that traveled on the transport's loading ramp. Another Hercules carried the drums which contained the special U-2 fuel. Under cover of darkness, the C-130s would fly to the deployment base, followed at an appropriate interval by the U-2 itself.[14]

The C-130s were loaded and dispatched to Peshawar airbase in northern Pakistan, arriving there in radio silence during the early evening, when most of the base personnel had left. Under cover of darkness, the U-2A was ferried from Incirlik by Frank Powers, landing just after midnight. The B-camera was loaded in the darkened hangar, while mission pilot Marty Knutson was briefed on the flight and began prebreathing. At six the following morning, before most of the airbase had woken up, Knutson took off on Mission 4125.

It was an almost complete success, although the flight was detected periodically by the PVO. Knutson eventually departed denied territory to the south as planned, and approached the Iranian base. He was relieved to see the C-130 which had brought the recovery crew parked below. Knutson had half-expected to find them under attack from the mountain bandits, having been told at his preflight briefing about the concern over security there. After nine hours ten minutes in the air, the U-2 was virtually running on fumes. After Knutson landed, they found only 20 gallons left in the sump tanks. The aircraft was quickly unloaded, refueled, and flown back to Incirlik by Barry Baker.[15]

The imagery of Tyuratam from Mission 4125 revealed preparations for what turned out to be a second successful Luna rocket launch, nine days later. Because of the pace of recent launches from here, the missile analysts had been expecting to see a second or even third launch pad in use already. The imagery *did* reveal a second launch complex 14 miles northeast of the first one, but it was still under construction. The surprised analysts noted that "the single launch facility coupled with the relatively short periods between several launchings indicates a highly efficient approach to check-out and launch of large missiles."[16] The upcoming second launch complex was judged by CIA analysts to be a prototype for the sites at which the R-7 (now designated the SS-6 Sapwood by the West) would be deployed.

The analysts still badly wanted some imagery of those operational sites, wherever they had been built, or might be being constructed. Their best guess now was that the first SS-6 deployment was somewhere in the Polyarnyy Ural area (evidently, no new intelligence pointing to Plesetsk, as the first location had been received).[17] But although the USSR did not protest the 9 July flight, Eisenhower was not encouraged enough by its success to allow further flights. Especially after Premier Khrushchev accepted, in mid-July, an invitation to visit the U.S. in late September.

The U-2C Deploys

At North Base, two more of the CIA's U-2A aircraft received the J75 engine and joined the U-2C flight test program. The CIA pilots who were resident with the "headquarters" unit there also began flying the new model. They found the U-2C to be quite a handful, especially at high altitude and heavy weights when slipper tanks were installed. This was because of the aircraft's reduced longitudinal stability, compared with the U-2A. Lockheed conceded that the U-2C could not be flown manually for extended periods of time. Accordingly, the entire autopilot adjustment box was transferred to the cockpit. This allowed the pilot to "tweak" the pitch, roll, and yaw characteristics of the autopilot for better performance while inflight.

For a time, the J75 showed a tendency to engine roughness and flameouts when climbing between 40,000 and 60,000 feet. Just like the bad old days of the early J57-P-37 powerplant! The problem was solved on an interim basis by reducing power, and eventually by adjusting the fuel control, but not before CIA pilot Jim Cherbonneaux set a new record of eight flameouts on a single flight!

After 400 hours of flight tests in only 11 weeks, the C-model was declared ready for deployment on 1 August. On 12 August, Articles 351 and 358 departed Edwards bound for Det B in Turkey. A six-man team from the Skunk Works followed, to check out the pilots and maintenance crews at the Det on the re-engined aircraft. Following the by-now regular routine for ferry flights, the aircraft made refueling stops at Plattsburgh AFB, NY, and Giebelstadt. By the end of August, ten pilots had been checked out at Det B on the re-engined bird: five CIA, four RAF, and the recently-appointed detachment commander, Colonel William Shelton. He was a newcomer to the U-2 operation, and a replacement for Stan Beerli, who was posted to DPD in Washington as Director of Operations.

In September 1959, another two U-2As were re-engined with the J75, test-flown, and ferried to Det C in Japan. On 24 September, Tom Crull made a familiarization flight from Atsugi in one of them, Article 360. Having experienced for himself the C-model's improved high-altitude performance, Crull ran short of fuel on his way back. A descent from 75,000 feet in the U-2C would consume 35 gallons and take 20 minutes, but the first 10,000 feet took a long time. The pilot had to ease the throttle to idle and extend the gear and speed brakes. Even then, the Angel hardly seemed inclined to leave the heavens; there was precious little drag created by the gear and brakes at these altitudes, and the engine thrust was still considerable even on minimum power. To cut it back further would invite a flameout. At the same time, care had to be taken to ensure that the speed didn't build up so that mach buffet was encountered. In view of all this, and the possibility of encountering strong headwinds during the descent, most pilots did not feel comfortable leaving cruise altitude with less than 100 gallons remaining.

Crull later said that, on this occasion, the gust control did not work, which slowed his descent even further. Whatever the reason, Article 360 ran out of fuel on approach to Atsugi when ten miles short of the runway. Fortunately, Crull had enough altitude remaining to deadstick the aircraft onto a small civilian airfield at nearby Fujisawa, although he was unable to lower the landing gear in time. In less than a minute, Crull was surrounded by curious Japanese civilians, some of them carrying cameras. He remained in the cockpit until help arrived from Atsugi, in the form of an L-20 Beaver lightplane carrying the detachment's security officers. They cordoned off the area and ordered the Japanese onlookers away at gunpoint. This only excited their curiosity still further, since the U-2 was flying completely unmarked save for a minute three-digit serial on the tail. Photographers snapped away, and the story reached the local papers. They asked pointed questions as to why a so-called weather research plane should cause such a security flap, and be flying in an all-black scheme without national insignia or registration marks! Article 360 was dismantled and returned to the U.S. for repair.

Meanwhile, at North Base, the U-2C prototype was now engaged on trials of "the Granger." This was the first item of electronic countermeasures (ECM) equipment to be added to the U-2, and it was designed to combat the new Soviet fighter radar. None of the ECM devices which had been developed for U.S. bomber and attack aircraft would fit on the U-2. The CIA let a contract to a small Palo Alto, California, company named Granger Associates. The task was to build a radar jammer which was small enough to fit in the brake parachute compartment at the base of the tail. (The parachute was rarely used, and was therefore dispensable).

The "Granger Box" used deceptive techniques, repeating the conical scan signal that was generated by an X-band fighter radar, but returning it 180-degrees out-of-phase. The radar jammer weighed only 38 lbs including the radome, but of necessity, its radiated power was small, and it was only effective in the rear quadrant. Still, it was judged to be better than no protection at all. Flight tests were completed in the fall of 1959, and all of the CIA's aircraft were quickly modified to accept the box.

Crowflight Re-organized

In September 1959, the 4080th Strategic Reconnaissance Wing began a new phase of Operation Crowflight. The grinding routine of twice-weekly, eight-hour flights up and down the sampling corridors had continued, although neither the U.S. nor the USSR had conducted a nuclear test in the atmosphere since November 1958, when a mutually-agreed moratorium came into effect. The filter papers of the aircraft's A-Foil (nose) and P-2 (equipment bay) systems were still collecting plenty of debris particles, however. But the Defense Atomic Support Agency (DASA, which replaced the Air Force Special Weapons Project in a 1959 re-organization) wanted to cover the northern latitudes more comprehensively, as nuclear scientists continued their quest to understand the long-term dispersal pattern of nuclear fall-out. In August 1959, therefore, det 4 at Ezeiza was temporarily closed, as well as det 3 at Ramey, and all six "hard-nose" sampling U-2As returned to Laughlin. The following month, three of them were sent to a new det 9 at Minot AFB, ND. The other three formed det 10, which remained at Laughlin, but made a deployment to Ramey every five or six weeks.

SAC was asked to extend the U-2 sampling flights from Minot all the way to the North Pole, but the 4080th wasn't ready for this yet. Apart from the challenges of grid navigation, and flying a single-engine aircraft over such hostile territory, there were also problems in determining in advance the winds that the flights would encounter in the upper atmosphere. In the lower latitudes, this was accomplished by taking data from radiosonde balloons sent aloft by weather stations. There were none of these stations in the polar regions. DASA eventually arranged three B-52 sampling flights to the North Pole instead.

Meanwhile, the furthest north that SAC's U-2s would venture was about 75 degrees, and this was only on the Toy Soldier missions out of Eielson AB. The "customer" for these flights, where the primary purpose was to diagnose the nature of Soviet nuclear warheads, was renamed AFTAC—the Air Force Technical Applications Center, in September 1959. The 4080th wing mounted its fifth deployment to Alaska in late September, for three weeks.

Back at Laughlin AFB, the 4080th continued to train for wartime and other contingencies with the 20 other U-2A models on strength there. SAC did not re-engine any of its aircraft, and they did not receive the Granger box or the System 6 ELINT upgrade. The wing did keep four aircraft in a dedicated SIGINT configuration, with Systems 1/3/4 loaded, but they were proving troublesome. The aircraft's rotating anticollision light interfered with System 1, and new filters had to be designed. The mean-time-between failure for System 3 was only eight hours. Moreover, the RADAN/ASN-6 system which had been added to these "ferret" aircraft for more precise navigation never worked properly. In mid-1959, it was replaced by a VOR/ADF installation, but still these SIGINT U-2s flying with SAC had not yet made an operational deployment.[18] SAC was also persevering with the Westinghouse APQ-56 SLAR on three U-2As, though this too had never been used for a real, intelligence-gathering flight. In September 1959, the 4080th failed its Unit Simulated Combat Mission, a test of operational readiness.

The 4080th wing gave up most of its Martin RB-57Ds to USAFE and ARDC in mid-1959; the twin-engined reconnaissance aircraft, which SAC had once championed over the CIA's U-2, had never been a great success. In particular, the Martin company's two main attempts to shave weight from the design had both caused serious problems after the RB-57D entered service. The honeycomb wing panels delaminated from the thin metal skin which covered them, and were a maintenance nightmare. The reduced-thickness wingspar failed on one aircraft as it landed, causing all aircraft to be grounded and then structurally strengthened.

The 4080th retained just six RB-57D2 models, a dedicated ELINT aircraft in which a second crew member was carried to operate a complex, semi-automated "ferret" system designed by Martin. During a deployment to Eielson in mid-1959, one of these aircraft confirmed that the latest Soviet early-warning and height-finding radars were capable of detecting aircraft at least 200 nautical miles beyond their Far Eastern borders.[19]

Khrushchev Keeps Quiet

Premier Khrushchev paid his visit to the U.S. in the second half of September as planned. Despite many opportunities to raise the subject in private, he never mentioned the U.S. spyplane overflights to President Eisenhower. Soon after he had gone, the CIA was back in the Oval Office trying to get permission for the long-mooted U-2 flight over northern Russia. But Ike turned it down.[20] A limited penetration over the Soviet Kurile Islands to the northeast of Japan was flown instead, on 1 November 1959.

In late August, the ad-hoc panel which had been advising the CIA on Soviet ICBM progress reminded Allen Dulles that the U-2 "possesses altitude capabilities which make it a unique platform for the reliable acquisition of high quality telemetry data prior to first stage burnout on Tyuratam ICBM launchings. Such data is of extreme importance in determining ICBM characteristics."[21] From September to December 1959, Det B mounted another 14 TELINT flights along the Soviet southern border in connection with missile test-launches from Tyuratam. A new dedicated missile telemetry system was added to the detachment's newly-converted C-models.[22] On some flights, pilots were even given a 16mm handheld Bolex movie camera, to film the missile's plume as it ascended from the launch pad if they could.[23] Some of the Soviet missile firings took place at night, and even though the U-2 was flying 500 miles to the south, the pilots witnessed the sky light up for hundred of miles.

First "British" Overflight

The pace of missile testing evidently encouraged the President to approve another mission to the launch sites in November, and the overflight fell to a British pilot for the first time. Since arriving at Det B almost a year earlier, the four RAF pilots had taken their turn to fly the SIGINT and TELINT flights along the border, as well as the continuing flights over the Middle East. Det B had even made three deployments to the UK, in December 1958 and May and October 1959, to show the U-2 to the RAF's top brass. Codenamed Operation HIGH TEA, these deployments were conducted as operational tests of Stan Beerli's Quickmove procedure. Each time, a C-130 and a U-2 were flown to RAF Watton in East Anglia. "Genuine" U-2 weather reconnaissance missions around the UK and out to the Bay of Biscay were flown from Watton by the RAF pilots.[24] (The British had elaborated on the U-2 cover story for their own purposes. Officially, the four RAF pilots were temporarily assigned to the Meteorological Office in London as "Experimental Officers").

However, Dick Bissell's original scheme whereby the British would stage their "own" flights over the USSR had not worked out in practice. Eisenhower had never really been persuaded that the U.S. would somehow be absolved if a U-2 was brought down over the USSR with a British pilot at the controls. So no separate chain of approval to London, bypassing the White House, had ever come into play.[25] But when the President approved the new November mission, the flight plan was transmitted by secure means to the small office within the Air Ministry in London which managed the RAF's U-2 involvement. From there, Squadron Leader Colin Kunkler took the plan across Whitehall to 10 Downing Street, accompanied by the assistant Chief of the Air Staff (Operations), Air Vice Marshall John Grandy. "The Americans want us to fly here, and here," Kunkler told the British Prime Minister, as he laid out the route map on a table. As sanguine and unruffled as ever, Harold Macmillan had no objection. Mission 8005 was on.[26]

It was launched from Peshawar on 6 December 1959 and flown by Wing Commander Robbie Robinson, the leader of the British group within Det B. The targets included Tyuratam and Kyshtym again, and Kapustin Yar, the IRBM test-launch base on the Volga River which had not been overflown by a U-2 since September 1957. In the intervening period, the big GE radar and the listening posts in Turkey had monitored continuing and frequent missile tests from there. Further up the Volga River, Robinson flew over the bomber production plant at Kuybyshev, and his route was also designed to cover as many railroads in these areas as possible, in the continuing search for an operational Soviet ICBM base.

At the end of the mission, Robinson was due to "coast-out" of denied territory over the Black Sea and recover directly into Det B's home base at Incirlik. But the Det was worried about the reduced range of the U-2C compared with the U-2A. In case the RAF pilot was low on fuel and thus obliged to land at one of the airbases in northern Turkey, the Det launched a recovery crew in a C-130 complete with ferry pilot Jim Barnes. Robinson was supposed to contact Barnes in the C-130 on a prearranged frequency, but they never heard his call. In any case, the British pilot successfully returned to Incirlik.

No operational ICBM bases were found on the imagery from Robinson's flight, but the PIs did discover a line-up of Bison bombers at Engels airfield opposite Saratov, and many developments within the seven missile test launch complexes at Kapustin Yar. Since the USSR also trained its operational missile troops here, there was much to be learned from detailed study of this imagery, especially concerning the mobile medium-range R-12 nuclear missile (SS-4 *Sandal*). It was not yet operational, but imagery of the trailers, erectors, warhead vans, and other equipment now being tested here, would reveal the patterns to look for in future deployments. (Nearly three years later, the USSR would dispatch the R-12 and the R-14, its longer-range cousin, to Cuba, and thus provoke the Missile Crisis of October 1962).

Slow Soviet Progress
On 15 December 1959 the huge R-7 missile roared off the first launch pad to be completed at Angara (Plesetsk). This was indeed the first Soviet operational ICBM base, although no-one in Western intelligence yet knew this for certain. The USSR had taken 30 long and hard months to make the base ready. The long, hard winters followed by springtime thaws slowed the pace of construction, especially in the swampy wastelands which had been chosen for the site. The massive Tyulpan launch structure required a firm foundation, and the slimy mud had to be excavated to the underlying bedrock. Angered at the costly delays, Premier Khrushchev nearly canceled the project in 1958, in favor of other sites. The construction plan for Plesetsk was reduced from 12 launch pads to only four. Two days after the successful launch, Khrushchev announced the formation of the Strategic Missile Forces, a new branch of the armed services.[27]

The first launch of an Atlas D, the U.S. equivalent to the R-7, had taken place on 9 September 1959 from Vandenberg AFB in California. Seven weeks after this, SAC placed the first Atlas ICBM on alert there. The U.S. and the UK already had Thor and Jupiter IRBMs pointing at the USSR from Europe. But there was still no way of knowing whether the Soviets had deployed enough ICBMs to hold the U.S. at ransom, as the missile gap theorists feared. Also, had the Soviets reached the same decision as the U.S., to curtail production of the first-generation liquid-fueled ICBMs in favor of solid-fueled designs which could be placed in silos to reduce their vulnerability?

On 19 January 1960, the USSR demonstrated an extended range for the R-7 when a test-launch overflew the usual impact area on Kamchatka and landed near Johnson Island in the Pacific Ocean. That same month, the latest formal estimate of Soviet ICBM capabilities was finalized in Washington. NIE 11-8-59 admitted that the U.S. still had "no direct evidence of Soviet ICBM deployment concepts or of the intended nature of operational launch sites." Were they rail-mobile units, hard or soft fixed installations, or some combination of these methods? The analysts simply didn't know. The NIE tentatively suggested a total of 35 Soviet ICBMs on launchers by mid-1960, and 140-200 by mid-1961.

In a long footnote to the top-secret NIE, the USAF strongly disagreed with this estimate. Influenced by the hawkish views of General Curt LeMay and his successor as SAC commander, General Tommy Power, USAF intelligence reckoned the

Soviets were going all out to "force their will on the U.S. through the threat of destruction." Power made a public speech which suggested that "the Soviets could virtually wipe out our entire nuclear strike capability within 30 minutes." Senator Symington alleged that "the intelligence books have been juggled so that the budget books can be balanced." At the end of January, Allen Dulles countered the missile gap theorists with a public speech of his own.[28]

More Flights Urged

On 2 February 1960, the President's Board of Consultants on Foreign Intelligence Activities (PBCFIA) held another of its periodic meetings with Eisenhower in the White House. General Jimmy Doolittle and the others urged the President to use the U-2 over the USSR "to the maximum degree possible." In reply, Ike told them that this was one of the most soul-searching decisions to come before a President. He had opened a personal dialogue with Premier Khrushchev during their meetings at Camp David the previous September, and now a Summit Meeting of the big four powers (the U.S., UK, France, and USSR) was planned for mid-May. "If one of these aircraft were lost when we are engaged in apparently sincere deliberations, it could be put on display in Moscow and minimize the President's effectiveness," Eisenhower continued.[29]

It was a sentient observation. Nevertheless, by the time that he met the PBCFIA, the President had already approved another U-2 mission into the southern USSR, encouraged perhaps by the knowledge that the British would be sharing the political risk again, with an RAF pilot at the controls. Like the previous overflight, this would depart from Peshawar, Pakistan, and recover into Det B's home base at Incirlik, Turkey. Once again, the flight would search for missile sites along the railroads of the Urals and the Volga River valley.

Det B deployed the fuel, launch crew, and mission pilot (Flight Lieutenant John MacArthur) to Peshawar in the usual manner on 3 February. But when John Shinn, one of the Det's American pilots, tried to ferry the mission aircraft from Incirlik later that day, it malfunctioned enroute. It was Article 360, the aircraft that had belly-landed in Japan and since been repaired by Lockheed. Since arriving at Det B, 360 had acquired a bad reputation. Each of the hand-built U-2s had their own idiosyncrasies, but 360 was universally disliked by the pilots. For one thing, its autopilot never behaved properly.

There was a 24-hour delay while a substitute U-2C was prepared. But it arrived behind schedule at Peshawar. There was a huge rush in the dark to get it refueled and prepared. On the morning of 5 February, mission 8009 took off late. There was no time to adjust the celestial precomputations. MacArthur crossed the border heading northwest. The skies were crystal clear, but most of the ground below was covered in deep snow. Beyond the Aral Sea, the terrain below was virtually featureless, except for the Emba and Ural river valleys. At the flight's most northerly point, MacArthur flew over the Soviet Aircraft Factory No 22 at Kazan. This had been an objective on the previous British U-2 mission, but Robbie Robinson had cut short that portion of the flight when his aircraft unexpectedly began contrailing.

This time, though, the U-2 flew straight over the snow-covered airfield and captured on film eight aircraft of a new design with highly swept wings and long pointed noses. It was a new supersonic medium bomber which was later given the codename BLINDER. The CIA checked previous intelligence reports, and found that reliable observers had noted a new type of aircraft there in the summer of 1958. That must have been a prototype of the new bomber, the analysts concluded, and now they were looking at a development or pre-production batch. Kazan had previously produced the Tu-16 BADGER, and it was correctly assumed that this was another Tupolev design. (It wasn't until July 1961 that the USSR finally "revealed" the Tu-22 in public, at the Moscow airshow).[30]

From Kazan, MacArthur turned south and flew down the Volga, covering more airfields and following more railroad lines. Crossing the Donetsk basin, his last major target was the city of Dnepropetrovsk. Sergei Korolev's R-7 missiles were built here, though the CIA didn't yet know that for sure.[31] Moreover, this was the location where Mikhail Yangel had established a "rival" design bureau for offensive missiles. His first major project was the medium-range R-12, which had been observed at Kapustin Yar on the previous U-2 overflight. But Yangel was also working on a second-generation ICBM which would be far more operationally flexible than the R-7.

MacArthur turned south and eventually left Soviet airspace at Sevastopol in the Crimea. The commo section at Incirlik waited anxiously for his call. The flight had fallen even further behind schedule when MacArthur found himself off-course when approaching one of his targets. He made a wide 360-degree turn to ensure that he was properly lined up. Eventually, the British pilot landed at Incirlik, and the film was rushed to the U.S. by courier plane, as usual. When it reached the Stueart Building, Art Lundahl's photo-interpreters scoured the extensive imagery of railroads for any sign of a Soviet ICBM deployment. They found none whatsoever.[32]

CORONA and OXCART Problems

Despite his misgivings, President Eisenhower reviewed DPD's plans for four more U-2 overflights in mid-February.[33] There was no technical alternative. Since the first, unsuccessful launch of the Discoverer/CORONA reconnaissance satellite one year earlier, there had been no fewer than nine more failures. There had been problems with the Thor-Agena launch rocket; the ingenious but complicated panoramic camera; and the film capsule ejection and recovery system. Morale within the program was low, and the technical competence of the CIA (eg the Development Projects Division) to manage the project was called into question.[34]

Meanwhile, progress with the stealthy and high supersonic successor to the U-2 had been even slower. A joint DoD, USAF, and CIA selection panel chose Kelly Johnson's A-12 proposal over the rival entry from Convair labeled Kingfish in August 1959. But the A-12 design needed further refinement, with the CIA's Land Panel providing advice, notably from Ed Purcell on the aircraft's radar cross section. The project acquired the cryptonym OXCART, and DPD brought John Parangosky back from his job as Detachment B's executive officer to manage it. Finally, on 11 February 1960 the CIA issued a firm contract to the Skunk Works for 12 aircraft. But the first aircraft wouldn't be ready to fly until August 1961. Moreover, Johnson noted in his diary that the OXCART project was "12 times as hard as anything we have done before."

Notes:
[1] Russian air defense source. The radar was later assigned the codename Scan Odd, and the missile was designated AA-1 Alkali, by the West
[2] Lockheed Report SP-179, "Flight Test Development of the U-2C Airplane", 1 July 1960 declassified 1999

3 White House Office of the Staff Secretary, Memorandum of Conference with the President, 22 December 1958, by John D. Eisenhower

4 White House Office of the Staff Secretary, Memorandum for the Record, 12 February 1959, by John D. Eisenhower.

5 White House memo, 4 March 1959

6 Varfolomeyev, p263

7 A White House memorandum which records the 3 April discussion exists, but has never been declassified

8 these are approximate figures. For the exact data, see Appendices

9 Seymour Hersh, "The Samson Option", Random House, NY 1991, p52-58

10 the term 'in the barrel' derives from 'a filthy fighter pilot's joke', so the author was informed

11 Bruce Bailey, "A Crow's Story", unpublished manuscript 1992, posted on 55th SRW Association's website 1998, p5

12 Pedlow/Welzenbach, p162. These authors suggest that the other aircraft involved in this successful telemetry interception was a USAF RB-57D, rather than an EB-47TT. However, no RB-57D was deployed to the area at this time.

13 White House Office of the Staff Secretary, Memorandum for the Record, 7 July 1959; and Memorandum of Conference with the President, 8 July 1959, by Andrew Goodpaster

14 Beerli interview

15 Beerli, Knutson interviews.

16 Memo to the DCI from the Ad-Hoc Panel on the Status of the Soviet ICBM Program, 25 August 1959, declassified by the DDEL.

17 as above

18 4080th wing history, November/December 1959

19 4080th wing history, July 1959

20 information from White House memo, supplied by Dwayne Day

21 Memo to DCI from Ad-Hoc Panel

22 this is believed to have been designated System 7, and to have featured automatic intercept receivers that were activated when launch control frequencies were used, and high-speed tape recorders. It may have used an additional pair of antennas on the upper fuselage of the U-2, which were better tuned to receive the L-band telemetry transmissions from Soviet missiles

23 Ericson, Knutson interviews

24 One of the weather flights out of Watton didn't go according to plan, because the aircraft had a hydraulic failure over the Bay of Biscay, and had to divert into the USAF base at Brize Norton near Oxford.

25 Bissell book, p116-117, maintains that the separate chain of approval really did exist. This author can find no evidence of this, even after interviewing RAF officers who were involved with the U-2 project at the command level. See next footnote

26 author interviews with Wg Cdrs Norman Mackie and Colin Kunkler.

27 Zaloga, "Target America", p150-151

28 Jackson, volume 5, p87-90

29 White House memo by Goodpaster, 8 February 1960

30 RAF Air Intelligence Review, April 1962, declassified as file AIR40/2556 in the PRO

31 NIE 11-5-60, "Soviet Capabilities in Guided Missiles and Space Vehicles", approved 3 May 1960 noted that there was "limited evidence pointing to two Soviet cities as possible production sites" for the ICBM

32 Incidentally, although much of the terrain imaged by MacArthur's flight was covered in deep snow, that did not entirely negate the intelligence value of the imagery. As Dino Brugioni explained to a conference on the CORONA satellite program in 1995, photo-interpreters learned much from the snow clearance patterns adopted by the USSR. For instance, the paths and roads to the headquarters buildings of any given complex were always the first to be cleared!

33 Pedlow/Welzenbach, p167

34 Dwayne Day, "The Development and Improvement of the CORONA Satellite" p57, in "Eye in the Sky, The Story of the CORONA Spy Satellites," Smithsonian Institution Press, Washington , 1998

9

Going for the Grand Slam

The Ad-Hoc Requirements Committee (ARC)[1] chaired by Jim Reber now designated four areas of the northern USSR as the highest-priority targets for U-2 missions. They were the railway lines from Kotlas to Salekhard, Perm to Vologda, Vologda to Archangelsk, and Petrozavodsk to Pechenga. Almost anywhere along these 2,500 miles of rails, they thought that a rail spur leading to an operational ICBM site might be found.[2] At Project HQ, though, the question of how to cover these northern targets had exercised the mission planners time after time. The biggest problem was choosing launch and recovery bases which were within range of the targets. The problem was exacerbated by the political constraints which were imposed on U-2 operations, by those foreign countries whose bases were the best situated for takeoffs and landings.

For political reasons, it was now impossible to launch overflight missions out of Norway or Turkey.[3] The UK or West Germany could be persuaded, but their territory was too far from the main areas of interest inside the USSR. In any case, the depth of Soviet radar coverage along its European borders guaranteed that any flight from there would be detected at an early stage. Thanks to Ayub Khan, the military leader of Pakistan whom the CIA had assiduously cultivated, Peshawar was still available. But there was no way that a U-2 could fly from there to the railways in the northern USSR and back again.[4]

Two different solutions to the problem were proposed by the mission planners of Project CHALICE. They were codenamed Operations TIME STEP and GRAND SLAM. The former involved launching the U-2 out of the USAF base at Thule, way inside the Arctic Circle on Greenland's western coast. It would then fly across the Arctic Ocean and the Barents Sea, and cross into Soviet airspace over Novaya Zemlya Island. The nuclear test site there would be a bonus target, before the aircraft flew over the southern Kara Sea and crossed the Soviet mainland coast. Then it would turn southwest to begin the search of the railroads, passing over Kotlas before turning north and heading for Plesetsk. From there, the flight would continue to Severodvinsk, where the CIA's clandestine sources suggested that the shipyard was building nuclear submarines. From there, the flight would cross the White Sea and follow another railroad north to the Murmansk naval base. By now running short of fuel, the U-2 would then turn west and fly down the Norwegian coastline to a landing at Andoya or Bodo.[5]

Norway would not be explicitly informed that the arriving U-2 had just completed an overflight. There was another political risk attached to this flight. The launch base at Thule in Greenland was theoretically part of Danish territory. There were other disadvantages to Operation Time Step, not least the likelihood that the U-2 would wander off course during the long initial transit across the Arctic Ocean.

The second proposed overflight was equally ambitious, because it entailed flying all the way across the USSR from south to north. For this reason, no doubt, the mission was codenamed Operation GRAND SLAM. However, it had some advantages. Firstly, it could be launched from Peshawar with Pakistan's full approval. It could then fly over Tyuratam again, before heading north to that rich complex of targets in the Urals industrial heartland...Magnitogorsk, Chelyabinsk, Kyshtym, and Sverdlvosk. From there, the flight could follow railroad lines northwest to Perm, Kirov, and Kotlas. Then it would follow the same route as the proposed Operation TIME STEP, to a landing in northern Norway.

Range versus Altitude

Project HQ had been mulling over a flight straight across the USSR for many months; Bissell had mentioned the idea to President Eisenhower in July 1959, during the discussions about Operation TOUCHDOWN. Now that the re-engined U-2C was available, it was a more practical proposition, even though some range was sacrificed when maximum altitude was required. Operation GRAND SLAM would be a nine-hour flight covering about 3,300 nautical miles. By coincidence, this was exactly the range to zero fuel which Lockheed had calculated for the U-2C's maximum altitude profile. Clearly, this allowed no fuel margin at all, and was not an acceptable option for mission planning.

However, because the new J75 engine could be operated anywhere between minimum fuel flow and maximum power, various reduced power cruise profiles were possible with the U-2C. Each of these profiles would extend the range by improving the fuel consumption. In fact, the aircraft could theoretically fly for as long as 11 and a half hours and 4,600 nautical miles. But this maximum range profile required the pilot to enter a cruise-climb at only 57,000 feet. By so doing, the aircraft's fuel consumption in the second half of the flight would be a miserly 2.7 miles per gallon, but even by the end of it, the U-2 would be no higher than 68,000 feet. By now, the threat from Soviet air defenses was simply too great to risk so low a cruise.

The best compromise between altitude and range was to fly the first part of the mission in the maximum power profile, entering the cruise-climb at 67,000 feet as usual. After 1,300 nautical miles, the aircraft would reach 70,000 feet, whereupon the pilot would level off. By flying the rest of the mission at this altitude, the range was increased to 3,800 nautical miles, and the endurance to nine hours 45 minutes.[6]

It was, therefore, technically possible for the U-2C to reach the northern USSR on either of Operations TIME STEP and GRAND SLAM, with a recovery in Northern Norway. Meanwhile, though, two less ambitious flight plans were also proposed to President Eisenhower in mid-February. They covered similar territory to the previous three overflights. Both would launch out of Peshawar. Operation SUN SPOT would revisit Tyuratam and Kapustin Yar. Operation SQUARE DEAL would fly over the former but not the latter, ranging instead further to the east to get new and more complete coverage of the very large missile test range at Saryshagan and the Semipalatinsk nuclear test site.[7]

White House Meetings

How would the President react to the latest flight plans? Dick Bissell went to the White House again, accompanied by Dr. Herbert York, the Director of Defense Research and Engineering in the Pentagon. York was a nuclear scientist whom Eisenhower had come to know through the President's Scientific Advisory Council (PSAC), where York served alongside Land, Purcell, and the others. Like the PSAC chairman James Killian, York didn't believe in the missile gap. But he didn't entirely trust the CIA's estimates, either. Unless the intelligence on Soviet missiles could be improved, the U.S. would have to accelerate its strategic weapons programs. York supported the idea of more U-2 flights, especially over the northern USSR.[8] The weight of advice was now irresistible. Eisenhower conceded that one of the four proposed U-2 missions could be flown, with a deadline of 30 March.

Launching from Pakistan was the favorite option. The CIA still believed that missions staged from there had a good chance of evading the Soviet early-warning radar net. This was crucial, since, if the PVO did not initially detect the flight, they might not be able to alert radar, missile, and interceptor forces further in the interior in sufficient time to effect a shootdown. This was especially so, since the U-2 flightpaths often zig-zagged their way across denied territory, with frequent turns and only a few long, straight flight legs. This was done mainly so that multiple intelligence targets could all be overflown on a single flight, but the frequent turns also served to reduce the opposition's chances of tracking and intercepting the aircraft.

The trouble with Operation GRAND SLAM, in particular, was that by flying all the way across the USSR, the options to zig-zag the U-2's flight path were much reduced. This theoretically made the U-2 more vulnerable, if the Soviet air defenses had now perfected a means of interception.

The SAM Threat

Project HQ asked the USAF's Air Technical Intelligence Center (ATIC) for a new assessment of Soviet capabilities against the U-2. Dick Bissell's military deputy, Colonel Bill Burke, reported the findings to his boss on 14 March 1960. "The greatest threat to the U-2 is the Soviet SAM," Burke told Bissell. ATIC had concluded that "the SA-2 *Guideline* has a high probability of successful intercept at 70,000 feet providing that detection is made in sufficient time to alert the site."[9]

Since its first appearance in late 1957, the PVO had been steadily deploying Petr Grushin's new S-75 Dvina surface-to-air missile system to protect the major cities and military-industrial complexes of the USSR. It was capable of detecting and tracking a fighter-size target at about 110 miles, using the P-12M surveillance radar. Each S-75 regiment had one of these at the command post, and another at each missile battery[10] under its command. The target's range, altitude, and bearing was initially determined by the regiment, which then passed the data to the nearest individual battery, where another P-12 and the dedicated engagement radar would take over. The missiles themselves were improved through further testing at Saryshagan, with the slightly longer V750VK and V750M versions demonstrating greater range and being deployed in 1958-59. Each battery consisted of six missile launchers, and it was normal practice to salvo-fire three missiles at each target, in order to increase the probability-of-kill.[11]

But some Western analysts believed that the missile's control surfaces weren't big enough to provide adequate guidance in the thin upper atmosphere which was inhabited by the U-2. That view was reflected in the latest National Intelligence Estimate on Soviet air defenses. This NIE was being finalized at the very time that Bill Burke received the USAF technical assessment which DPD had commissioned. The NIE was approved for release just two weeks later, on 29 March 1960, and credited the S-75 with less high-altitude capability than the USAF report to DPD. The NIE suggested that the "maximum effective altitude (of the SA-2) is about 60,000 feet, with *some* capability up to about 80,000 feet, especially if employed with a nuclear warhead."[12] The NIE hedged its bets on whether such a warhead was yet available to the PVO (in fact it wasn't, and no nuclear warhead was ever planned for the S-75). But the NIE did state unambiguously that the SA-2 had been deployed rapidly and extensively "to defend major centers of population and industry." The NIE listed locations within the USSR where SA-2 sites had been identified, including Baku, Kharkov, Kiev, Moscow, Odessa, Rostov...and Sverdlovsk.[13]

The big city in the heart of the Urals was on the planned flight path of Operation Grand Slam. The SA-2 missiles had been deployed at Sverdlovsk since at least July 1959, when Vice-President Nixon made a goodwill tour of the USSR. Nixon and his party visited Leningrad, Novosibirsk, and Sverdlovsk, as well as Moscow. Before they left Washington, they were briefed by Lundahl's experts in the PIC, on potential intelligence targets.[14] As the Soviet airliner carrying the Americans approached Sverdlovsk from the southwest, a young State Department officer noticed an unusual pattern of revetted clearings grouped around a radar trailer and linked by service roads. Ray Garthoff leaned towards the window and surreptitiously photographed the site.[15]

In June 1959, intelligence officers assigned to the British Mission to Soviet Forces in Germany (BRIXMIS) learned that the new Soviet SAM had been deployed at Glau, just south of Berlin. They couldn't get close enough to photograph it from the ground. So they took to the air, flying the mission's Chipmunk, a two-seat lightplane which was based at RAF Gatow. They trespassed beyond the Berlin Air Safety Zone and flew towards the site, staying at treetop height to evade detection. Wielding his camera, the rear-seat passenger got excellent close-ups of the missile site, including the revetted launchers, the *Spoon Rest* surveillance radar, and the *Fruit Set (Fan Song)* missile guidance radar.[16]

BRIXMIS had obtained the best intelligence yet on the new Soviet SAM. Later in the year, ELINT intercepts of the radars at the Glau site were obtained. The missile analysts reviewed some earlier data: previously-unidentified ELINT intercepts; the 1957 U-2 imagery of the Lake Baikal test range; and the *Guideline* missile's first appearance in Red Square later that same year. All the intelligence pointed to a sustained deployment of the new SAM since 1957.[17]

The U-2 itself brought back more imagery of the SA-2 in December 1959, when Robbie Robinson flew over Kapustin Yar. There were *Guideline* missiles protecting the missile test complex, and more of them were found in another area, where PVO troops were being trained to fire them.[18]

In sum, Western analysts had made a slow start on assessing the new Soviet SAM, but they were now catching up with the nature and the extent of the threat posed by the SA-2. But the missile analysts apparently did *not* know one vital piece of information, which might have caused them to take the SA-2 threat even more seriously. Grushin's S-75 system had *already* shot down a high-flying enemy spyplane!

However, the action had not taken place over Soviet territory. It was over mainland China instead. On 7 October 1959, Captain Wang Ying Chin of the Republic of China Air Force (ROCAF) on Taiwan flew a Martin RB-57D reconnaissance aircraft deep into communist mainland China. As we have seen, the RB-57D was a poor relation to the U-2 in the spyplane business, but it could reach over 60,000 feet. When SAC decided to phase out the Martin design, two of the 4080th wing's machines were assigned to a joint USAF-ROCAF project to overfly the People's Republic.

Unknown to the U.S. or Taiwan, the People's Republic of China had secretly received five SA-2 systems from the Soviet Union in late 1958. After a period of training, which evidently was not affected by the growing Sino-Soviet split, the missiles were declared fully operational in defense of Beijing during late September 1959. But no public announcement was made. Just over a week later, Captain Wang's flight took off from Taiwan. The *Fan Song* radar from one of the SA-2 batteries identified his aircraft from 70 miles as it flew towards the communist capital. Three missiles were fired at 43 miles range, and the RB-57D was shot down some 18 km south of the capital. Beijing announced that it had intercepted and ended the spyflight, but was careful not to specify the exact means. Wang was listed missing, believed killed. The USAF intelligence officers who were assigned to the ROCAF were unable to determine for certain what had happened. They presumed that fighters had intercepted the RB-57D.

In fact, this incident near Beijing was the world's first downing of an aircraft by a surface-to-air missile. Moreover, Wang had been flying at 63,000 feet when he was shot down.[19]

The SA-2 missile's maximum range against a high-flying B-52 was estimated to be 25 nautical miles. Mission planners in Project HQ had the option of routing the U-2 clear of the engagement zones around SA-2 sites—if they knew where the sites were actually situated. But if the sites were defending prime intelligence targets, such as Tyuratam, that would defeat the object of the flight. In reality, therefore, the U-2 mission planners hoped that the PVO's early warning system was still not good enough to provide adequate data to the missile sites. But if it *was* good enough, they hoped that the aircraft's high cruising altitude would provide a last but vital margin of defense.

In view of the now-crucial importance of avoiding detection by the Soviet early warning radars, the mission which was planned to approach the northern USSR across the Arctic Ocean (Operation TIME STEP) was now rated the least desirable. Burke told Bissell on 14 March that this flight stood a 90 percent chance of being detected on entry. If so, it could then be tracked accurately throughout the four hours that it would spend in denied territory. The flight would evoke "a strong PVO reaction (including) alerting of SAM sites and pre-positioning of missile-equipped fighters in the Murmansk area (point of exit), thus enhancing the possibility of successful intercept." Even if the Soviets didn't manage an intercept, Burke continued, they would have enough radar data "to document a diplomatic protest." He recommended Operation SQUARE DEAL out of Pakistan for the next overflight, because there was "a reasonable chance of completing (it) without detection."[20]

This was an optimistic assessment, because the USSR had now virtually completed its early-warning net along the southern border. Indeed, the new NIE on Soviet air defenses issued two weeks later warned of the "impressive" extent of the cover provided by the increased numbers of *Token* and *Bar Lock* radars, with detection ranges of 190-220 nautical miles and from 70,000 feet (*Token*) to 220,000 feet (*Bar Lock*). "Gaps in (Sino-Soviet) peripheral early warning radar coverage now appear only in southwestern China," the NIE also noted.[21]

More SAC photo flights

The steady improvement in Soviet air defenses was bad news for SAC as well. In the event of war, would the nuclear-armed B-47 and B-52 bombers get through? One area of the USSR which was of particular interest to SAC's war planners was the Siberian coastline, since many of the U.S. bombers would fly over the North Pole or across the Bering Sea to attack their assigned targets in the Soviet interior. The Generals in Omaha decided to send their own U-2s to probe the Soviet defenses around Siberia. The 4080th wing was alerted to deploy to Alaska for a second series of peripheral reconnaissance missions codenamed CONGO MAIDEN. Like the first series exactly one year earlier, these would be long, challenging, and potentially dangerous flights over icy and remote waters. But, although the aircraft would come within range of Soviet air defenses, SAC didn't have to get political authority from Washington for the flights. As long as the pilots managed to fly a few miles off the Soviet coast, they would be legal.[22]

The 4080th pilots practiced for the new deployment by flying 15 miles off the Texas coast. They put a notch in the hand control of the U-2's driftsight, to help them maintain the correct distance, and honed their celestial navigation skills. Where they would soon be flying, there were few radio stations to help them navigate by ADF! Major Joe Jackson was selected to command the detachment. On 11 March 1960, three aircraft left Laughlin for Operating Location Five at Eielson AB, ready to fly the photo missions now that the dark northern winter was coming to an end. As in the previous CONGO MAIDEN deployment in 1959, the aircraft were configured with the B-camera, plus System 1 for ELINT and System 3 for COMINT on any reaction from the Soviet air defenses.[23]

On 15 March, Majors "Snake" Bedford and John McElveen both flew successful photo missions from Eielson along the Soviet Bering Sea coastline. A few days later, Major Ed Dixon covered the Chukotsky Peninsula as far as Mys Schmidya and flew around Wrangel Island. Major Dick Leavitt was assigned the next flight, the very long haul across the ice-covered north coast to Tiksi and back. He had flown the mission profile a number of times from Laughlin, to ensure that the U-2A had sufficient range. After waiting some days for a good weather report from SAC HQ, Leavitt took off. He flew a great circle route over the Arctic and intercepted the Siberian coast near Ambarchik. The sun was still very low and straight ahead, making it difficult to identify the pressure ridges, and to distinguish the islands or the coastline in the eternal whiteness below. Eventually, he saw Tiksi off to the right and realized that he had accidentally strayed over the mainland! He quickly turned right, flew past the remote Arctic coastal city, and headed northeast. After rounding the New Siberian Islands, Leavitt headed back to Alaska.[24]

There was no doubt that these Arctic flights by the USAF's blue-suiters were detected by the opposition. The USSR had already sited early warning radars on some of their northern offshore islands, as well as the mainland coast, to defend against SAC's bombers. The PVO couldn't protect the entire coastline with fighters, but the SAC U-2 pilots saw contrails beneath them occasionally. On one of the flights, the System 3 COMINT recorder brought back an excited conversation between a GCI controller and the pilot of a Yak-28 fighter which was attempting to intercept the U-2. Through such SIGINT, plus imagery of the coastal radars and airfields, however, SAC's intelligence experts could re-assess what chances the bombers would have of penetrating. The USSR did not protest Leavitt's unintentional overflight. SAC planned more U-2 missions in the Arctic.

Deadlines Approach

In Washington meanwhile, the deadline of 30 March which the President had specified when approving one of the CIA's four proposed overflights was fast-approaching. There had been a delay in gaining permission to use Peshawar, and no deployment had yet been mounted. On 28 March, Eisenhower agreed to extend the deadline to 10 April, and even authorized a second overflight to be performed by 19 April.[25] On 1 April, the CIA station chief in Oslo sent a request to the Norwegian government, requesting permission to use Andoya airbase for two U-2 flights during April. Andoya was a remote island 140 miles further north even than Bodo, and hence that much closer to the Soviet border from which the U-2 would be coming, if either of the two options for a northern overflight were exercised. The CIA told Colonel Wilhelm Evang, Norwegian intelligence chief, that the purpose of the U-2 flights was SIGINT. The Norwegians were preoccupied with a large NATO naval exercise in Northern waters, and suggested a postponement until at least 19 April. They also recommended Bodo instead of Andoya, because it offered greater security.

The northern flight could therefore not now be launched before the President's latest deadline. But weather conditions were improving over the southern USSR. Project HQ decided to launch Operation SQUARE DEAL—the flight to Tyuratam, Saryshagan, and Semipalatinsk. Det B was alerted, and on 8 April the usual routine for an overflight deployment swung into action. It was Bob Ericson's turn "in the barrel," but he was accompanied aboard the C-130 on the trip to Pakistan by the next-in-line U-2 pilot, Frank Powers. He would be a back-up, in case Ericson fell sick at the last minute. The overflights were now so few and each so critical that no chances were being taken. The back-up pilot would attend the preflight briefing and go though the prebreathing process, just in case.

Operation SQUARE DEAL

The deployment to Peshawar went smoothly, and the U-2 was ferried in on schedule later that night. In the early morning sunlight of 9 April, Ericson took off on Mission 4155. He headed north to cross the Soviet border over the Pamir range of mountains. But only 150 miles into Soviet territory, and contrary to the mission planners' best hopes, the PVO detected the flight. The Turkestan Air Defense Corps had radars dotted all over the Pamir range. The U-2 was shielded from the first of these at Khorog by high ground. But stable tracking of the aircraft was accomplished by a second site at Kara-Kul to the north. Col-Gen Yuri Votinsev had revitalized the poorly-organized Turkestan Corps since being given its command a year earlier. He was in the command post, and ordered four MiG-19 fighters to take off from Andizhan airbase. They were accurately vectored towards the target by ground controllers. But the MiGs could only reach 52,500 feet, and failed to make contact.[26]

System 6, the ELINT detector on the U-2, picked up the Soviet tracking. But there was no real-time display to the pilot of the data, which was all recorded on tape for subsequent analysis on the ground. Ericson pressed on towards his first major target, which was the missile test range at Saryshagan, on the shores of Lake Balkhash.

The next objective was 400 miles to the north: the strategic bomber base at Dolon and Semipalatinsk nuclear test site. Here, the flight plan called for Ericson to fly a search pattern so that maximum coverage of the huge test area was obtained. Unknown to the U-2 pilot, two of the latest Sukhoi Su-9 interceptors were on alert at the nearest PVO base, some distance away. The supersonic tailed-delta fighter had been rushed into service with the PVO before the formal state acceptance process was completed. The powerful Su-9 was equipped with four beam-riding K-5 air-air missiles. But it was short on range.

The Su-9 pilots calculated their fuel and realized that, if they were to reach the U-2's position, they would have to land afterwards at the airfield on the nuclear test site, rather than return to their own base. But that airfield was off-limits to anyone without a special top-secret clearance. The correct procedures had to be followed! Frantic messages were passed up the PVO's chain of command from the regiment to the air defense district to the headquarters in Moscow. Here, the duty officer woke Marshall Biryuzov, who informed Defense Minister Marshall Malinovsky, who telephoned the Minister of Medium Machine Building, who controlled the Semipalatinsk site. Meanwhile, the new Sukhoi jets stayed firmly on the ground. By the time the required clearance was issued, Ericson had completed his search pattern and was headed south out of range.[27]

The U-2 headed back towards Saryshagan, to photograph another part of that large air defense site. This was quite a risk, since test and training firings of the S-75 SAM system were regularly conducted here. In fact, the latest version of the *Fan Song* engagement radar was currently being tested. It had a greater ability to hold target track at high altitude. Although no test flights were scheduled for this day, two missiles were ready on one of the launch pads. However, instead of warheads, they had miss-distance indicators fitted for the test flights. As the U-2 flew back over the test range, the test site commanders did contemplate firing the missiles, despite the lack of warheads, but decided against doing so because the chance of a successful interception was virtually nil.[28] Unaware of the commotion below, Ericson set course for Tyuratam and flew blithely on.

A pair of MiG-19s from the 356th Fighter Regiment at Sverdlovsk were assigned to the chase. This was one of the few PVO units whose MiGs were fitted with the new RP-5 search radar. The pilots were kept on alert, in their pressure suits. This regiment had shot down some of the MOBY DICK balloons that the U.S. attempted to float across the northern USSR in mid-1958. But the two MiGs had to stop at Omsk for refueling, and they failed to arrive in time. Another pair of Su-9 interceptors were also available in this area. But they belonged to a new unit which was converting from the MiG-19 and wasn't yet combat-ready. Missiles from the MiG-19 were hurriedly loaded onto one of the new jets while the other took off. Inexperienced GCI controllers were unable to vector it towards the target. The second Su-9 also took off. Its pilot, Captain Darashenko, did catch a glimpse of the U-2 from 57,000 feet, but he was inexperienced in zoom climb techniques, and couldn't gain the extra 10,000 feet required to bring the missiles to bear. The interceptors returned to base, but one of the MiG-19s crashed on approach to Sverdlovsk, killing the pilot, Vladimir Korchevsky.[29]

From Tyuratam, Ericson turned south and headed for the border. By now, he had been over Soviet territory for nearly six hours. Occasionally, he had seen the fighter contrails below as they vainly tried to intercept. During most of the overflight, Marshall Sergei Biryuzov had maintained a grim, silent vigil at the PVO's Moscow headquarters, while the catalogue of PVO errors unfolded. The U-2 was now back in Col-Gen Votinsev's area of command. As it neared Mary in the southern Turkmen SSR, Votintsev sent a message to Biryuzov, suggesting that the intruder might start his descent once he had left Soviet airspace. In desperation, two MiG-19s were scrambled from Mary with orders to chase the U-2 over the border, and destroy it by ramming if necessary. Shadowing the Angel above them, the MiGs flew nearly 200 miles into Iran on Ericson's tail. But the U-2's destination was the Iranian airstrip at Zahedan, 500 miles south of the border, and it did not descend yet. The MiGs returned to Mary with their fuel nearly exhausted. The U-2 landed unscathed, and the Lockheed recovery crew soon had it turned round and flown back to Turkey.

Premier Khrushchev had to be informed, and he was furious at the PVO's failure. The chastened Biryuzov set up an enquiry. It was another exercise in self-denial. Anastas Mikoyan, co-leader of the MiG design bureau, told the enquiry that no aircraft in the world could fly for more than six hours at 20,000 metres. This was despite Andrei Tupolev's continuing assertion that it was indeed possible (the famous bomber designer had sketched out the design of such a plane for Khrushchev three years earlier). In typical Soviet style, most of the blame was placed on the local air defense commanders for their "criminal lack of concern," and the enquiry called for them to be severely punished for failing to bring down the flight. However, Biryuzov was also criticized by his political masters for failing to anticipate where the spy flights were going, and concentrating his forces there. "Do not lose heart," Biryuzov told Votintsev, "in air defense, one who has been flogged is worth not two, but a dozen who have not. Remember this!"[30] The entire PVO was put on a heightened state of alert.

Bonanza

Ericson's flight was another intelligence bonanza for the CIA's photo-interpreters and analysts. It returned the first useful imagery of Tyuratam since Marty Knutson's mission nine months earlier, and revealed a third launch complex. This consisted of two pads, but they were being constructed in a different manner from those at the first two complexes, from where the R-7 ICBM and its derivatives for the Sputnik and Luna space projects had been fired.[31] This was the first indication to U.S. intelligence that the USSR might be preparing to test Yangel's second-generation Soviet ICBM.

Indeed, engine tests for the new R-16 ICBM had begun in late 1959, and the prototype missiles were moving off the assembly line at Dnepropetrovsk. The R-16 still used liquid fuel, but it was smaller than the R-7 and with no strap-on boosters. Because it used hypergolic fuel with nitric acid as the oxidiser, the R-16 could be more easily fueled, and it remained ready for launch for up to two days. The "Big Seven," on the other hand, had to be laboriously fueled with kerosene, plus the liquid oxygen which acted as the oxidant. This was a 20-hour process, and if the missile wasn't fired almost immediately, the R-7 then had to be de-fueled before the super-cold liquid oxygen reverted to its natural gaseous state.[32]

At Saryshagan, Mission 4155's B-camera captured for the first time two new radars which were soon associated with Soviet attempts to create an antiballistic missile system. It was a significant find; the CIA had not estimated that the Soviets were serious about ABM development until two months earlier, when Marshall Konev made a tantalizing reference to the possibility in a speech to Warsaw Pact Defense Ministers. The technical approach that the USSR was taking to ABM was completely unknown. The analysts made a further scrutiny of the buildings surrounding the two radars, which they codenamed *Hen House* and *Hen Roost*. They concluded that ABM research and development was a high priority in the USSR, and had probably been going on at Saryshagan since early 1959.[33]

Premier Khrushchev decided not to send a protest note about the 9 April flight. "Why give our enemies the satisfaction?" he told his son Sergei. According to the latter, the Soviet Premier did not want a confrontation. He thought that Eisenhower had not personally authorized the flight, and calculated that when he heard of it, the American President would certainly not approve another.[34] The USSR had not protested the previous overflights in July and December 1959 and February 1960, either.[35]

This time, though, the CIA had unambiguous evidence that the latest flight had been detected at an early stage. It came from the read-out of System 6,[36] and also possibly from U.S. ground stations along the Soviet border which intercepted the PVO's communications. In his postflight report, Ericson also detailed how he had seen fighter contrails beneath him.

Unfortunately, Bissell and his subordinates in Project CHALICE failed to draw the appropriate conclusion from Mission 4155—that time had finally caught up with the U-2. Moreover, while the impressive intelligence gains from the flight were becoming known in mid-April, Project HQ was already in possession of that most rare commodity—Presidential authority for a further overflight!

Temptation

The temptation to use that authority was just too great. It is not clear whether Eisenhower was informed of the flap in the Soviet air defenses which had been caused by the 9 April flight. However, there was apparently a meeting in the White House in the middle of April when Bissell reviewed the overflight situation for the President, in the presence of Secretary of State Herter and Hugh Cumming, who ran the State's Bureau of Intelligence and Research.[37] Bissell explained that the delay requested by Norway, plus too much cloud over the target areas, had prevented the second, northern overflight mission before the President's new deadline of 19 April.[38] He asked for more time. Eisenhower had imposed the deadline because the Paris summit of the Big Four powers was now looming. On Monday 25 April, Goodpaster called Bissell to tell him that Eisenhower had extended the deadline to 1 May. "No operation is to be carried out after May 1," Goodpaster emphasized.[39] The summit was due to open on 16 May.

Which operation was it to be? Project HQ had by now discounted Operation TIME STEP out of Greenland and across the Arctic Ocean because of the air defense threat. That left Operation GRAND SLAM out of Pakistan, the one that flew right across the USSR, to cover the unknown territory in the north. But on Tuesday 26 April, Burke warned Bissell that "penetration without detection from the (Pakistan) area may not be as easy in the future as heretofore."[40] Nevertheless, the go-ahead was given.

Burke and his staff knew how delicate the situation would be at the recovery base in Norway. Operations chief Colonel Stan Beerli was dispatched from Washington to take charge of operations there. Beerli had commanded the October 1958 deployment to Bodo, and so already knew the key Norwegians who would have to be involved. At the same time, Det B was alerted to dispatch the launch and recovery crews from Incirlik. On Wednesday morning 27 April, a C-124 transport was loaded with 55-gallon drums containing the special U-2 fuel, and took off for Pakistan. A C-130 Hercules followed carrying Det B commander Colonel William Shelton, the launch crew, the mission pilot, and the backup pilot. The pilots were, respectively, Frank Powers and Bob Ericson, two of the project's most experienced and well-regarded "drivers." Two more C-130s left Incirlik with the recovery crew, including Marty Knutson, the pilot who had been assigned to fly the U-2 back to Turkey as soon as the mission was over. These C-130s would stopover in Germany until instructed to fly on to Norway.[41]

After a seven-hour flight from Turkey, including a refueling stop in Bahrain, the C-130 carrying the launch crew arrived at Peshawar. The C-124 was already there. The 20-strong party unloaded all the equipment and fuel from the two transports into the hangar, which was some way from the main part of the base. Meanwhile, Stan Beerli arrived in Oslo and renewed his acquaintance with Colonel Wilhelm Evang, the Norwegian intelligence chief. He told Evang that the proposed "SIGINT operation" would take place within the next 24 hours.[42]

Glen Dunaway took off from Incirlik for Peshawar to ferry the U-2C which would fly the mission. It was Article 358, the best plane of the four that Det B possessed. Night fell as Dunaway flew across Iran and Afghanistan. He arrived at the Pakistani base around 2 am. Meanwhile, Powers had been trying to sleep on a camp bed in the hot and noisy hangar, without much success. He had just been woken and was dressing when the commo people got word relayed from Washington that the mission had been postponed 24 hours. There was too much cloud over the target areas.

Shelton told Ericson to fly the U-2 back to Turkey. This was a departure from the previous routine. On recent operational deployments, if a mission was postponed, the spyplane was kept hidden in the hangar. This time, though, it would be sent back to Turkey and flown out again the next night. Ericson took off before daybreak for the five-hour return flight to Incirlik. Glen Dunaway took Ericson's place as the mission back-up pilot to Powers.[43]

Weather Factor

In Washington at 1717 H Street, mission planners scanned the weather reports anxiously. Most of the information on which they based a go or no-go decision was provided by the Global Weather Center at SAC headquarters in Omaha and the USAF weather forecasting group at Andrews AFB. But some of it came from routine Soviet weather forecasts, which were broadcast to all-comers! Operation GRAND SLAM had particularly demanding weather criteria, since the maximum success would only be achieved if the route all the way across the USSR was cloud-free. The probability of such conditions was not great! In fact, the whole of central and northern Russia was currently covered in cloud. There was even a late snowstorm predicted for the Urals.[44]

At Incirlik, John Shinn was assigned to ferry the U-2 back to Pakistan. Towards evening on Thursday 28 April, he flew off to the east. When he landed in the small hours at Peshawar, the mission was still on. Powers and Dunaway had finished breakfast and were "on the hose" when the word came from Washington: another 24-hour postponement! Dunaway completed his prebreathing, and flew the aircraft back to Turkey again. Shinn now took his place as the back-up pilot to Powers.

The contingent at Peshawar whiled away the long, tedious hours playing cards and reading. They weren't allowed to leave the hangar or its immediate surroundings. They cooked their own food from rations they had brought with them on the C-130. Some slept on camp beds. On Friday afternoon, just before Powers and Shinn were due to go to bed again, a further word was passed from the commo van. No flight on Saturday, either.

Accidents and Incidents

Frank Powers, therefore, had plenty of time to reflect on the unusual nature of his postponed mission. He was probably more aware than most of his colleagues about what could go wrong. Because he was the safety officer for Det B, Powers received all the accident reports concerning the U-2.

There had been no fatalities or airframe losses in the U-2 program since September 1958. But there had been plenty of incidents, some of them serious. Within the CIA's detachments, Marty Knutson had suffered a complete electrical failure during a border SIGINT flight at night. With his faceplate partly frosted over, Knutson had somehow recovered the aircraft to Incirlik. He flew nearly four hours using a flashlight to illuminate the sextant display, through which he identified the North Star which indicated roughly which way he should fly! Robbie Robinson had stalled an aircraft and lost 20,000 feet trying to recover, after deploying the gust control without first raising the nose. Sammy Snyder was returning from an overflight of China in 1959 when the bearings froze up. He managed to dead-stick the aircraft into Taoyuan airfield, Taiwan.

The SAC U-2 wing also reported plenty of incidents. Captain Pat Halloran had seen the low-fuel warning light come on during a climbout: the fuel wouldn't feed from the wing tanks due to a maintenance error. Halloran turned back, but the engine flamed out and he was forced to make a silent return to the runway through heavy overcast. A trainee pilot suffered a broken fuel line during a long-distance navigation leg at night, forcing him to make a deadstick landing at a remote civilian airport in Colorado. A tailpipe adapter collapsed as one aircraft throttled up for takeoff, causing an explosion.[45]

Most recently, two U-2s had made forced landings on a frozen lake and in a rice paddy. The frozen lake was in Saskatchewan, Canada, where Captain Roger Cooper of the 4080th wing had been forced down on 15 March 1960 after the aircraft's battery malfunctioned and then blew up. Although the cockpit filled with smoke and the engine quit, Cooper descended through a 10,000 feet overcast to make a textbook deadstick landing. He had been flying a sampling mission out of Minot AFB, ND. The aircraft was later flown off the lakebed.

The rice paddy was in Northern Thailand. CIA pilot Bill McMurry had ended up there on 5 April after running short of fuel during another overflight by Det C of southern China and Tibet. The flight originated at Cubi Point in the Philippines, and was due to land at Takhli in Thailand. Two similar missions had been successfully accomplished seven days earlier. But unknown to McMurray, his tailwheel doors failed to close after takeoff. On such a long flight, the additional drag took a steady toll of the aircraft's fuel reserves. McMurray apparently didn't pay close enough attention to his green card, a standard takealong on each U-2 flight, on which the mission navigator predicted the fuel remaining against elapsed time. McMurray didn't notice the extra fuel consumption until well into the flight. When the low-fuel warning light came on, he was obliged to penetrate the solid overcast and search for a clearing in the jungle below. He managed to crash-land the aircraft. Working up to their waists in water, a recovery crew dismantled it with the help of local villagers.

How long would it be before a technical failure took place during a Soviet overflight? There was no doubt that the CIA detachments enjoyed superior maintenance by the contract Lockheed employees, and every possible technical contingency was covered in preflight planning. But there was such a thing as The Law of Averages! And nine-hour missions really stretched the endurance of both man and machine.[46]

Incidents over Siberia

Ed Dixon knew this only too well. On 16 April, SAC had launched him on another of the long flights in the CONGO MAIDEN series along the Soviet Siberian coast. He was repeating Dick Leavitt's flight of three weeks earlier, because SAC wanted better imagery. Above the ice-covered wilderness, Dixon also had difficulty in staying 15 miles offshore. Moreover, as he turned into Tiksi Bay, he saw a flash on the ground, as if a missile had fired. Dixon hurriedly turned the autopilot off and hand-flew a 180-degree turn to quickly exit the area. He decided to return the way he had come, through the narrow Laptev Straight, rather than fly round the New Siberian Islands. Even so, he ran short of fuel on the way back to Eielson. The warmer temperatures above the northern tropopause meant that the U-2 went a little faster and used more fuel. Dixon squeaked into Eielson with 22 gallons remaining.[47]

Around the same time, another CONGO MAIDEN flight was passing the coastal town of Anadyr on the Chukotsky when an SA-2 was fired at it. The pilot was blissfully unaware at the time, and the attack wasn't discovered until the imagery from the flight was processed and analyzed. The missile's characteristic shape ("like a long telegraph pole") appeared to rise and then level off some way below the U-

2. This seems to have been the first confirmed occasion on which the Soviet air defenses fired an SA-2 at a U-2.[48] The PVO had never fired one at any of the CIA's penetrating overflights. But the PVO was about to get another opportunity.

Still Waiting...

Powers and the launch crew whiled away Friday night and all day Saturday in Pakistan. They didn't have the big picture, of course, to understand why there was such a hold-up. Apart from Shelton and a few key people in Washington, no-one did. The Agency's compartmented security procedures took care of that. The commo people had the best understanding, since they sent out all the messages with their top-secret routing slugs and codenamed addressees.

On Saturday morning 30 April in Project HQ, the duty team checked the weather reports yet again. It was still cloudy over the southern USSR, but the bad weather had cleared from the north. It was now or never! The White House had stipulated that no mission be flown *after* 1 May. Dick Bissell was away for the weekend. Duty staffers contacted DCI Dulles himself, for final authority to launch the flight. After consulting with CIA Deputy Director Pierre Cabell, Dulles gave the go-ahead.[49]

At Incirlik, Det B had a problem. Because the mission aircraft had been ferried to Pakistan and back twice, it had run out of hours. The 200-hour phase maintenance check was now due. So a different aircraft was substituted. It was Article 360, hardly the pilots' favorite. On Saturday evening, Bob Ericson flew it to Peshawar.

He arrived about three am. While the groundcrew readied the aircraft yet again, the pilots went into the makeshift briefing room. Shelton reviewed the mission again. Powers thought it looked a long way. He worried that in case of mechanical trouble, there were no short escape routes out of hostile airspace. Ericson simply didn't believe the plane could go the distance. There was a long discussion between the pilots and Shelton about this. The Det B commander suggested that, if Powers was running short of fuel when he reached Kandalaksha on the Kola Peninsula, he could abandon the remaining targets, turn west and head straight for Norway across Finland and Sweden. Powers and backup pilot Shinn started suiting-up. Ericson would act as "mobile," helping Powers strap in and taxi-out.[50]

Colonel Shelton asked Powers if he wanted to take the poison pin. Since the first overflights in 1956, pilots had been offered a means of committing suicide, in case of capture and torture. Until recently, the means had been a small cyanide capsule.[51] After Shelton joined Det B, he pointed out that there would be unforeseen consequences, if the capsule should accidentally break or leak in the pocket of the pilot's coveralls. With the help of the Agency's clandestine specialists, Jim Cunningham came up with an alternative. The Administrative Director at Project HQ provided a silver dollar with a tiny hole, into which was placed a needle smeared with a deadly toxin. Powers decided to take the silver dollar along.

The ground crew loaded the B-camera and the SIGINT systems. One of them mistakenly put Ericson's emergency seatpack back into the cockpit, instead of Powers'. The error wasn't noticed. The refueling process began: a painstaking process of transferring the JP-TS from the 55-gallon drums. It was important that this specially-refined kerosene was allowed to settle, so that each tank could be topped up completely. Three months earlier, an aircraft had been hurriedly refueled for an overflight mission after arriving late at Peshawar. The fuel being transferred from the drums was warm, and some spilled out over the wings, when the specified load couldn't be accomodated within the tanks. British pilot John MacArthur was seriously short of fuel by the end of the flight. Fuel temperature was an important variable. At Det C, they had even suggested that the U-2's fuel be deliberately precooled, in order to increase the aircraft's range. Powers remembered that the fuel system was one of Article 360's idiosyncracies; every now and then, one of the wing tanks failed to feed.

The scheduled takeoff time was 6 am. With 40 minutes to go, Powers moved out to the plane and Ericson helped him strap in and hook up to the radio, oxygen system, and so on. The PSD technician made his final check, and Ericson did the walkaround. The sun had already been up an hour, and Ericson took off his shirt to use as a makeshift sunshield over the cockpit. Ericson was still worried about the length of the mission. He reviewed the flight maps that Powers was taking, and made sure that the Finnish airfield of Sodankyla was clearly marked. Shelton had briefed this small airstrip as an emergency landing site, in extremis, if Powers was really short of fuel by then.[52]

They were all ready to go. The launch crew packed up the maintenance and PSD trailers and loaded them into the C-130, ready for the return trip to Turkey. The final signal to execute the mission had still not been received. This "go-code" always came all the way from Project HQ in Washington, along the CIA's own secure lines in Morse Code. Usually, it was sent four hours before scheduled launch time. This time, though, six o'clock came, and no signal. The commo van was still outside, set up some way from the plane. Inside, the operators hopped from one prearranged frequency to another but heard nothing, although there was a Morse signal on the guard frequency as they swept by it. Eventually, one of them tuned into guard, and realized that it was the transmission they had been seeking. The "go-code" had reached as far as Incirlik, but the Agency's radio operator there had been unable to transmit it onwards on the pre-allocated frequencies, because of the ionospheric interference which sometimes occurs at sunrise in springtime.[53]

Shelton had been standing anxiously behind the operators. As soon as the go-code was recognized, he rushed out of the cabin and ran towards the plane, signaling permission to depart. Powers had been in the hot cockpit for a whole hour, and his long underwear was soaked with perspiration. Ericson closed the canopy, Powers started up, and taxied the short distance to the runway. He pushed the throttle forward for takeoff and, as usual, the U-2 leapt into the air after little more than a thousand feet of runway. Mission 4154 was 26 minutes late taking off.

Notes:

[1] the ARC was renamed the Committee on Overhead Reconnaissance (COMOR) in mid-1960

[2] memo from COMOR, "List of Highest Priority Targets", 18 August 1960, in CIA Corona documents, p49-52

[3] it is not clear whether the US ever had permission from Turkey for a U-2 overflight to recover *directly* into Incirlik. Zahedan in Iran was the destination for the last two successful U-2 overflights of the USSR flown by US pilots. However, the British overflight missions 8005 and 8009 did recover directly to Incirlik, and this may have been a calculated political risk, especially considering that the UK had broken off diplomatic relations with Turkey at the time! That situation itself had caused an additional problem for Det B at Incirlik, where the British contingent had to maintain a *very* low profile!

[4] author's interview with Col Stan Beerli

[5] Pedlow/Welzenbach, p172

[6] Lockheed Skunk Works report SP-179, "Flight Test Development of the Lockheed U-2C Airplane," 1 July 1960

[7] Pedlow/Welzenbach, p170-171

[8] Herbert York, "Making Weapons, Talking Peace", Basic Books, New York 1987, p123, and author's correspondence with York

[9] Pedlow/Welzenbach, p168

[10] For clarity, I have used the word 'battery' to describe the individual firing sites of the SA-2 system, rather than the Soviet term 'battalion'. Confusingly, the Russian word for a battalion is 'diviziony'.

The Russian word 'diviziya' describes the next-highest level of command, namely the regimental headquarters!

[11] Zaloga, "Soviet Air Defense", p39-47; Col-Gen S. Mironov, "Several Questions on Evaluating the Effectiveness of the Basic Means of AntiAir Defense of a Front, and its Organizational Structure," Voyennaya Mysl (Military Thought), circa 1960, p7. Amongst the readers of this top-secret Soviet military journal was Col Oleg Penkovsky, who passed a copy of this and many other articles to CIA, whose translation is used here.

[12] author's italics

[13] NIE 11-3-60, "Sino-Soviet Air Defense Capabilities Through Mid-1965", 29 March 1960, declassified 1996

[14] Michael Beschloss, "MayDay", Harper and Row, New York, 1986, p178

[15] Ray Garthoff, comments at CIA U-2 conference, Washington DC, September 1998

[16] Tony Geraghty, "BRIXMIS", Harper Collins, London 1996 p90

[17] NIE 11-5-59, "Soviet Capabilities in Guided Missiles and Space Vehicles" approved 3 November 1959, declassified

[18] CIA Corona documents, p193

[19] details of Chinese shootdown (related here in English for the first time) are from "Dangdai Zhongguo Kongjun" eg History of the Chinese Air Force, published in the China Today series, China Aviation Industry Press, Beijing, 1989. Confirmation that US intelligence did not know that a SAM had shot down the U-2 is from NIE 11-3-60 dated 29 March 1960, p12 ("there is no evidence that any surface-to-air missiles have been given to any other Soviet Bloc nation"). It was probably the Soviet spy Penkovskiy who finally tipped off the West about Soviet missiles to China. During the course of an extensive grilling on the S-75 in 1961, he told his interrogators from MI6 and CIA that the Chinese had been given "all the types of guided rocket that are in production."

[20] Pedlow/Welzenbach, p170, 174

[21] NIE 11-3-60, p10,12

[22] exactly what that distance should be, was disputed. The US held that sovereign airspace extended only three miles from the coast, but the USSR insisted it was 12 miles. It wasn't until 1988 that the US unequivocally accepted the 12-mile limit.

[23] 4080th wing history, May 1960, Dixon notes

[24] Leavitt notes

[25] Pedlow/Welzenbach, p168, 170, 172. Mysteriously, White House records recording these decisions have not been found or declassified. In correspondence with the author, Herbert York recalls learning in later years, of Eisenhower's request that no notes be taken of these most sensitive discussions in spring 1960. The authors of the CIA's history used references to these White House meetings, which were contained within the individual U-2 mission folders. These folders still exist - but have not been declassified.

[26] Votintsev article, part 2

[27] Mikhailov and Orlov article, Khrushchev book

[28] Khrushchev book. Mikhailov and Orlov have an alternative version of this story. They say that the missiles themselves were not actually at the launch pad, that troops rushed to move them there from the storage depot, but did not get there in time.

[29] Mikhailov and Orlov article; notes provided to author by Yefim Gordon

[30] Votintsev article, part 2; CIA transcript of a debriefing of Col Oleg Penkovsky by CIA and MI6, 27 April 1961, paragraph 85

[31] CIA Corona volume, p195

[32] Zaloga, "Target America", p151-157, 194

[33] see, for example, NIE 11-4-60 "Main Trends in Soviet Capabilities and Policies 1960-1965", approved 1 December 1960, declassified

[34] Khrushchev book

[35] Mikhailov and Orlov article, implies that both of Det B's 1959 overflights were detected.

[36] Pedlow/Welzenbach, p170

[37] Leonard Moseley, "Dulles - A Biography of Eleanor, Allen and John Foster Dulles and their Family Network", Hodder & Staughton, London 1978, p453. Bissell, in memoirs p125, does not recall meet-

ing Eisenhower personally at this time. He says that Andrew Goodpaster relayed his request for an extension to the overflight deadline to the President

[38] Pedlow/Welzenbach, p172

[39] White House memo, 25 April 1960

[40] Pedlow/Welzenbach, p170

[41] The authors reconstruction of the sequence of events which led up to the launch of the final U-2 overflight is based on many sources. They include Powers book and prison diary, Tamnes book, Pedlow/Welzenbach CIA history, and interviews with Bob Ericson, Stan Beerli, Glen Dunaway, John Shinn, and Jim Woods. Where appropriate, travel orders and pilot's flight records have also been consulted

[42] The investigation which Norwegian authorities conducted following the abortive U-2 flight into Bodo suggests that Beerli told Evang that plans for the US operation had changed, and would now only involve C-130 flights. See Tamnes, p178. The USAF did indeed operate C-130s in Europe for SIGINT purposes. However, it cannot be discounted that Evang did not tell the whole truth to the investigation, about what he knew of the U-2 operation, and when he knew it.

[43] The author has not been able to determine who made the change of plan. Stan Beerli, the operations chief in Project HQ, says he knew of no such plan to ferry the aircraft to and fro before he left Washington for Norway on 26 April. The decision may have been taken by Colonel Shelton of Det B himself, although Colonel Bill Burke in project HQ should also have been aware

[44] According to testimony given by DDCI Gen Pierre Cabell to Congress after the Powers shootdown, meteorological records showed that there were only six days in the entire April-September period when weather conditions were ideal for aerial photography across such a large area of the USSR

[45] author's interviews, plus 4080th wing histories, August 1959, March/April 1960

[46] Skunk Works engineers will, to this day, point out that even when zero gallons were indicated on the fuel counter, there was a further 45 of usable fuel remaining - author interview with Bob Klinger

[47] 4080th history, May 1960, Dixon notes

[48] Dick Leavitt interview. Powers, in book p67, alleges that he and others in the U-2 program knew that SAMs were being fired at the later CIA U-2 overflights eg in 1959-60. However, the author has spoken the pilots of all four successful overflights in those years, and none recalls such an event. In evidence to the Senate Foreign Relations Committee on 31 May 1960, DDCI General Pierre Cabell quickly and unequivocally corrected DCI Allen Dulles, when the latter suggested that missiles had been fired at earlier overflights. It's not clear whether the intelligence derived from the imagery from SAC's U-2 flight past Anadyr, eg that a SAM had been fired, was available to the CIA project before Powers' mission was launched

[49] Allen Dulles' testimony to Senate Committee on Foreign Affairs, Executive Session, 31 May 1960

[50] Powers prison diary, Ericson interview. (In his autobiography written in 1970, Powers wrote that he knew the details of his route over the USSR some days earlier, and that he had complete trust in the aircraft. However, the author has preferred to rely on the diary that Powers wrote while imprisoned in the USSR. This was written six months after the event, not ten years later)

[51] On a flight over Bulgaria in December 1956, Det A Carmine Vito nearly mistook the cyanide capsule for a lemon drop. Vito occasionally removed his faceplate temporarily during a long mission, to suck one of his favorite sweets, although this was against orders. Fumbling for one of the sweets in his coveralls, he discovered that the ground crew had slipped the cyanide capsule into the same pocket. According to Pedlow/Welzenbach, p125, Vito didn't discover his mistake until he actually sucked on the capsule. However, Vito has assured this author that this story has grown with the telling, and the capsule never actually reached his mouth before he discovered what it was!

[52] After this fact emerged at Powers' trial in Moscow, the neutral Finns were embarrassed - and rather puzzled. Sodankylas airfield had been destroyed by the Germans during the Second World War. All that remained was a 2,500-feet gravel runway, which would have been a tough landing for the U-2!

[53] Pedlow/Welzenbach, p176. According to Powers (book), Shelton told him that the delay was because they were awaiting approval from the White House. This was not correct, although Shelton may have said this, for reasons unknown. As Allen Dulles made clear in his testimony to the Senate Foreign Relations Committee on 31 May 1960, he personally took the decision to proceed, "after consultation with (DDCI) General Cabell and other qualified advisors"

10

Poisoned CHALICE

Powers climbed skywards from Peshawar in the usual spectacular manner of a U-2. The roar of the J75 in the still morning air startled Pakistani airmen all over the base. Reaching the planned penetration altitude of 67,000 feet, Powers entered the cruise-climb and made the single-click signal to base on the UHF radio, to indicate that he was proceeding. Ericson acknowledged with another click. From now on, there would be no radio contact until Powers was ready to descend into Bodo nearly nine hours later. After 30 minutes in the air, the U-2 approached the Soviet border in the same Pamir region as the previous U-2 mission three weeks earlier. Looking ahead, Powers could see a solid undercast stretching from the lee of the mountains on the Soviet side and away into the distance. He cursed the late takeoff. All the celestial computations would be out, and he would have to resort to dead reckoning as the primary navigation aid.

The PVO's early warning radars soon detected the intruder, just as they had the previous overflight. The mission was easily plotted heading north, and the command post at PVO headquarters in Moscow was alerted. In the Soviet capital, the time there was one hour behind Pakistan. By 6 am Moscow time, Biryuzov and all the top PVO generals had been woken and rushed to the command post, which was in the Defense Ministry on the bank of the Moscow River. Over at the Kremlin, preparations for the MayDay parade were well-advanced. It was due to start at 10 am. The entire political and military hierarchy were supposed to be there, including Biryuzov and his men.[1]

Defense Minister Marshall Malinovsky woke Premier Khrushchev and told him about the new intruder. The Soviet leader ordered that it be destroyed regardless of cost. All air traffic across the entire Soviet Union was grounded, so that radar operators could concentrate on the vital target. On the plotting screen in the PVO command post, Powers' plane was moved steadily north. Interceptors were scrambled from Tashkent in pairs. As usual, they couldn't climb high enough to get near.

Powers had been over denied territory for well over an hour before he saw a break in the clouds. He spotted the Aral Sea off to the left in the distance, and realized he was about 30 miles off course. While he was making a correction, he noticed a contrail beneath him for the first time, moving at supersonic speed in the opposite direction. More interceptors had been launched from Tyuratam, which Powers was now approaching. But the pilots here didn't even have pressure suits, and they didn't make contact. Powers continued to monitor them through the driftsight, disappointed that the flight had been detected so early, but relieved that the fighters remained so far below.

There were thunderclouds over the launch site. Powers switched on the B-camera as planned, but noted that the intelligence would be less than 100%. There were three S-75 batteries at the missile site, but one was unmanned because of the May-day holiday, and the U-2 did not fly within the engagement zone of the other two. Now the solid undercast resumed. Powers tried a sextant shot, as a cross-check on the compass. It seemed to be accurate.

Pressure in Moscow

The PVO commanders in Moscow now expected the U-2 to turn northeast or northwest, to cover the same objectives as the three previous intrusions. But it headed straight on north. Biryuzov was fielding angry phone calls from Malinovsky, and even from the Kremlin. "It's a scandal," Khrushchev berated him. "The country gave all the necessary resources to the Troops of the Air Defense, yet you cannot destroy a subsonic plane!"[2]

It wasn't that easy, though. The PVO was still learning how to operate the new missiles, and how to properly co-ordinate all the radars involved. It took nearly a day to dismantle, move, and set up an S-75 battery. Once established, it took up to five hours to bring the battery to combat readiness. Even then, the diesel generators couldn't be kept running indefinitely. If there was no other source of power, they had to be turned on. That caused another 13-minute delay. If the target aircraft was flying at high subsonic speed, it had to be detected 80 miles away, in order to slew the missiles, acquire the target with the *Fan Song* engagement radar, and fire. But the P-12M radar, which began the search and acquisition sequence, had a range of only 110 miles against a fighter-size target. There wasn't much scope for error or delay. Furthermore, each battery could only remain at the highest readiness state for 25 minutes.[3]

By 0800 local time the U-2 had passed Magnitogorsk, and was clearly heading for the central Urals. Here were many nuclear weapons-related facilities which the USSR had taken great trouble to isolate from prying eyes in so-called "secret cities." Unlike further south, the skies over the Urals were fine and clear—great weather for aerial reconnaissance!

Autopilot Problem

Powers saw the clouds thinning below him. He peered through the driftsight, trying without success to relate the emerging ground features to his maps. Fortunately, his flight plan showed a radio beacon in this area, so he tuned the ADF. The specified frequency was correct, though the callsign was wrong. The radio fix helped him re-establish the correct course. He was some way south of Chelyabinsk. But with one problem solved, another one emerged. The autopilot malfunctioned, causing the nose to pitch-up. Powers retrimmed, flew manually for a while, then re-engaged the autopilot. The plane flew fine for a while, then pitched up again. Powers re-peated the corrective action, with no better result. Faced with the prospect of flying manually for another six hours, he nearly turned back. After all, the U-2C Flight Handbook clearly stated that "proper autopilot operation is mandatory...to satisfac-torily accomplish the mission."[4]

"An hour earlier, the decision to turn back would have been automatic," Pow-ers later recalled. "But I was more than 1,300 miles inside Russia, and the visibility ahead looked excellent. I decided to go on and accomplish what I had set out to do."[5]

An S-75 regiment had been located at Chelyabinsk. It was brought to highest alert status. A P-12 radar acquired the U-2, but the "hand-off" (transfer) of target from radar to missile battery went wrong. One of the radar circuits at the battery fused, and the missile control officer searched his screen in vain.[6]

A Sukhoi-9 interceptor was thrown into the hunt. The pilot took off and flew south towards the approaching target. But he had been scrambled too early: run-ning short of fuel, he was obliged to descend and land at the small airfield at Troitsk. This didn't even have a runway; it was the first time anyone had landed a Su-9 on grass!

Powers flew on, crossing unscathed over Chelyabinsk. He was relieved to see no interceptors beneath him. The next target was the USSR's first plutonium pro-duction complex, which was located amongst the lakes and forests to the east of Kyshtym. Powers was now flying over the scene of a serious nuclear accident three years earlier which the USSR had managed to hide from the world. High-level nuclear waste that was kept in solution within an underwater concrete tank had exploded and released 70-80 metric tons of radioactive debris into the atmosphere. Over 200 towns and villages with a combined population of 270,000 had to be evacuated. Despite the massive scale of this disaster, only the vaguest reports had yet reached Western intelligence (and it would be another 29 years before the full story emerged from behind the Iron Curtain).[7]

More Interceptors Launch

Powers turned onto a northeasterly heading which would take him towards Sverdlovsk. The PVO had now repositioned two MiG-19s belonging to the 356th Regiment from Perm to Koltsovo airbase, on the military side of Sverdlovsk airport. They were being hurriedly refueled. The same regiment had participated in the chase for Mission 4155 three weeks earlier. Another two Sukhoi Su-9s were also on the ramp. They were on a delivery flight from the factory in Novosibirsk to Belorussia, and had only stopped at Koltsovo to refuel. But the Su-9s had been held up there over the MayDay holiday, by the same poor weather which had caused Mission 4154 to be postponed three days running.

The pilot of the second Su-9, Captain Igor Mentyukov, had no pressure suit and the plane was unarmed. But Gen Yuri Vovk, who commanded the fighter-interceptors in the Sverdlovsk district, knew that the Su-9 had a better chance of reaching the high-flying target than the MiGs. He ordered Mentyukov to takeoff. As the Su-9 climbed away, Vovk radioed to Mentyukov: "There is a real target at high altitude. The mission is to destroy the target, to ram it...Dragon has ordered this." Dragon was the personal codename of Marshall Yevgeny Savitsky, the commander of all PVO fighter units, Vovk reminded Mentyukov.

Without a pressure suit to protect him if the cockpit depressurized, Mentyukov was facing almost certain death if he were to fulfill the mission. Under the usual tight control from the ground, Mentyukov was directed to turn towards the target. He jettisoned his drop tanks and accelerated to Mach 1.9. As he climbed through 65,000 feet, the ground controllers told him that the target was only 25 kilometers away. They could see both the interceptor and the intruder on their radar screens. But in the black sky all around him, the pilot could see nothing. Mentyukov switched on his radar, but the screen showed only interference. This was strange—it had worked just fine when he tested it just after takeoff.[8]

The supersonic Su-9 was closing fast on the subsonic U-2, and the controller ordered him to switch off the afterburner. Mentyukov protested that this would cost him altitude, but the controller repeated the instruction. The pilot complied, and the Su-9 began descending as it passed the still-unseen U-2. The ground controller now told Mentyukov he had overshot. After some mutual recrimination between pilot and controller, Mentyukov reported that he was running short on fuel. The controller told him to return and land, and the two MiG-19s were scrambled instead.[9]

The Missiles Engage

Meanwhile, Powers had reached a point some 35 miles southeast of Sverdlovsk. At the PVO district headquarters, the command post which controlled the missile batteries was on a separate floor to Gen Vovk's command post which directed the fighter interceptors. And the overall district commander was away on a training course. It was a recipe for confusion.

The missile command post identified the nearest missile battery to the approaching plane, and passed its range, altitude, and bearing to the site, with orders to fire. The commander of this battery was also away, and the task fell to his deputy, Major Mikhail Voronov. His three radar operators had to acquire the target by means of the two narrow fan-shaped beams generated by the *Fan Song*. It was a difficult task, since the U-2 was passing close to the edge of the battery's engagement area. Eventually, the operators managed to switch the radar into automatic tracking and missile guidance mode, and a firing solution was obtained. Voronov checked his own screen and ordered the launch, a salvo-fire of three missiles. The launch control officer hesitated, and Voronov yelled at him to fire. The solid fuel booster at the rear of one V-75 missile ignited, and it roared away. Another two missiles should have followed seconds later, but they didn't; the target was fast receding, and the fire control system calculated that it had now passed beyond range.[10]

But the first missile was streaking towards the U-2, some 14 miles distant. The command guidance link from Voronov's battery to the missile was properly established, and the missile accelerated past Mach 2. It would take only a minute to reach the target, even though the U-2 had now reached 70,000 feet. The fire control computer issued the appropriate changes of course as the missile closed on the target from behind. After flying for some 12.5 miles, the second-stage would be exhausted, and the missile would then continue on a ballistic trajectory.

Powers had no idea he was under attack. As the missile neared, he made a 90-degree left turn as shown on his flightplan, in order to line up on the next intelligence objectives on the southwestern fringes of Sverdlovsk. He was making routine notes of the time, altitude, speed, and EGT when "there was a dull thump, the aircraft jerked forward, and a tremendous orange flash lit the cockpit and sky," he recalled.[11] It was 0853 local time, and his long ambitious flight had been cut short after only three and a half hours.

Shoot-Down

The missile had exploded some way behind (and probably below) the U-2. The V-75 missile's warhead was designed to explode upon command guidance, or automatically when it reached close proximity to the target. A direct hit was not required, because the 180-kilogram fragmentation warhead broke up into 3,600 shotgun-type pellets, which were projected forward in an expanding cone.[12] The pellets hit the tail and rear fuselage of the U-2. In the cockpit, Powers was shielded from their forward path by the engine and wings. Maybe the sharp left turn which he had just made also helped the pilot escape the full force of the blast. Powers instinctively grasped the throttle with his left hand, and keeping his right hand on the big control yoke, corrected a slight drop in the right wing. But then the nose dropped, and no amount of pulling back on the yoke could bring it back up. Then a violent shaking erupted, flinging the pilot all over the cockpit. He later reckoned the wings had come off. "What was left of the plane started spinning, only upside down," Powers recalled. Experiencing up to 4g and with his pressure suit now inflated, the pilot had extreme difficulty trying to make his escape. He was being forced forward and up, and could not activate the ejection system for fear that his legs would be chopped off by the canopy rail when the rocket fired and the seat rose up its track.

As the spinning fuselage descended through 34,000 feet, Powers realized that he had an alternative means of escape—to bail out. He pulled the canopy release handle, and the canopy sailed away. He tried to reach the destruct switches on the right canopy rail, then changed his mind, thinking that he ought to ensure that he could escape before he operated them. So he pulled the "green apple" handle which activated his emergency oxygen supply, and released his seat belt. Immediately, the centrifugal force threw him halfway out of the aircraft, but he was restrained by the oxygen hoses, which he had forgotten to disconnect. His faceplate had now frozen over. He kicked and squirmed blindly, and suddenly he was free.[13]

On the ground, Major Voronov's crew saw the target dissolve into interference on their radar screens. They weren't sure whether this represented a successful kill, or whether the intruder had deployed electronic countermeasures against their radar, such as chaff. Voronov reported his uncertainty up the chain of command. The regiment commander Colonel Pevnyi ordered the adjacent S-75 battery at Beryozovsk, into whose zone of engagement Powers had now flown, to fire at the target. At this battery, Major Nikolai Sheludko and his crew successfully salvo-fired three missiles. They climbed towards what was now a descending mass of debris, comprising the U-2 fuselage, wings, and tail (all now separated), and the fragments of the single missile from Voronov's battery. Two of Sheludko's missiles either detonated close to the debris, or flew through it and detonated some way further on, as they were automatically programmed to do when no target had been intercepted (the third had flown off in completely the wrong direction).[14] By now, Powers had made his escape from the spinning, falling cockpit of the U-2. Once again, he was incredibly lucky. Somehow, he was not hit by the two newly-fired missiles which headed his way, or by the many pieces of aircraft and missile debris which were now falling from the sky.

PVO Confusion

That debris was now creating an even bigger bloom on the radar screens of the PVO's batteries and command post below. From a third S-75 battery to the north of the city, Major A. Shugayev reported that his crew was tracking targets which appeared to include the enemy aircraft, now descended to 36,000 feet. It was not responding to their IFF ("identification friend or foe") interrogation. There was still no outcome reported from Voronov's battery, or from Sheludko's. Pevnyi, the commander of the missile regiment, suspected that the targets which Shugayev had identified were the PVO's own interceptor fighters. But the district command in

Sverdlovsk didn't even realize that the fighter command post in the same building had some MiGs airborne. Some minutes passed, and still the target was being tracked at 36,000 feet. Pevnyi was ordered to engage the newly-reported target by the most senior PVO man present, deputy district commander Major General Solodovnikov. So Shugayev was cleared to fire, and another three missiles streaked skywards. A short time later, back at the first battery, Voronov finally realized that he had destroyed the intruder with the very first missile. He belatedly reported a confirmed kill.[15]

So what target had Shugayev's battery just fired at? Captain Boris Ayvazyan was leading the flight of two MiG-19s which had been scrambled from Koltsovo ten minutes before Powers was shot down. They were directed to fly to the north of the original missile engagement zone, towards Perm. The ground controller told Ayvazyan that the target was ahead of him by 10 kilometers, then only five. The pilot looked around for his wingman and found him some way behind, struggling to keep up. At the same moment, Ayvazyan saw an explosion above and to his right, and turned towards it. Was this his target? The flight leader wasn't sure (in fact, it was probably the first missile from Voronov's battery exploding and downing the U-2). Shortly after, Ayvazyan saw another explosion off to his right (probably one of Sheludko's missiles).

A third MiG-19 joined the pair. It was flown by the 356th Regiment commander Gennady Gustov, who had scrambled from their home base at Perm. He was having difficulty communicating with the ground controller, so Ayvazyan relayed his request for instructions. The answer came back: Gustov to return to Perm, Ayvazyan's pair to land back at Koltsovo and refuel.

Ayvazyan set course for the airfield and decided on a straight-in approach. His wingman was still some distance behind. Ayvazyan saw the runway ahead as he began to descend from 10 kilometers (33,000 feet). Suddenly, there was an urgent warning from the controller: "Lose height quickly!" Someone on the ground had realized that fighters and SAMs were now in the same firing zone, although they didn't explicitly warn Ayvazyan to this effect. Nevertheless, the flight leader reacted intuitively by putting his aircraft into a vertical dive, pulling out only 300 meters (1,000 feet) from the ground. His wingman Safronov was instructed to do the same thing, but no more was heard from him. Ayvazyan did a circuit of the airfield, hoping that his wingman would rejoin. But Senior Lt Sergei Safronov had been hit by one of the missiles fired from Sheludko's battery. He ejected from his stricken aircraft, and the parachute deployed, but the unfortunate pilot was found dead on the ground. The wreckage of the MiG fell into a public park near a village to the west of Sverdlovsk.[16]

According to the most detailed published account of the shootdown, at least one more missile was fired from a fourth battery. This one was successfully evaded by Igor Mentyukov, the Su-9 pilot who had been ordered to ram Powers' plane, as he descended to land at Koltsovo.[17] It was nearly 40 minutes after the shootdown before the PVO's Sverdlovsk command post finally realized that it had shot down the intruder, and stopped trying to track it![18]

Powers Captured

Powers saw nothing of this. Almost immediately after escaping from the aircraft, his parachute opened automatically. The American realized that he was already below 15,000 feet, where the barostat operated. He couldn't tell for sure, though, since his faceplate was still frosted over. He undid the clasp and removed it. He floated slowly down, just missed some power lines, and landed in a ploughed field near a village. Two farm workers helped him to his feet and removed the parachute. More people arrived. There was no chance of escape.

Confirmation that the spyplane was down was belatedly passed to Marshall Biryuzov in Moscow. He hurried to tell the Soviet leadership, which was already watching the Mayday parade. Western observers noticed a distraction as Khrushchev and others left the Kremlin reviewing stand temporarily, to hear Biryuzov's whispered report. A delegation of senior officers from the party, the KGB, the GRU, and the PVO was quickly dispatched to Sverdlovsk. They left Moscow's Vnukovo airport at midday in a Tu-104 airliner—the first aircraft allowed to take off since the nationwide grounding order had been issued. By the time that the airliner arrived at Sverdlovsk, Powers had already undergone his first interrogation, at the hands of the local KGB. It was decided to bring him secretly to Moscow. By evening, he was inside the Lubyanka Prison at KGB headquarters in the capital. The Soviet leadership was uncertain whether to publicize the shootdown.[19]

At Bodo airfield, Norway, the recovery crew waited in vain. They had already spent three nights waiting at Rhein-Main airbase, Frankfurt, for clearance from Washington to proceed to Norway. During the fourth night, the clearance had finally come. The two C-130s departed for Bodo early on Sunday morning, 1 May. One of them flew via Oslo to pick up Stan Beerli. Arriving at Bodo, Beerli met the most senior Norwegian officer there, Maj Gen Tufte Johnson. He arranged for the recovery team set up in the same remote hangar as they had used for the long deployment to Bodo in fall 1958.

Powers was due to arrive around midday local time. With an hour to go, Beerli told ferry pilot Marty Knutson to start pre-breathing. That way, the U-2 could be turned around and on its way back to Turkey as fast as possible. Minimum exposure! But the recovery crew listened in vain for the radio callsign "Puppy 68," which Powers was supposed to use as he neared the airfield. They waited five hours before giving up all hope and flying back to Oslo in the C-130. Using the CIA's secure line from the U.S. embassy, Beerli sent word to Washington that the plane was overdue.

At project HQ, they already suspected as much. As usual, the National Security Agency (NSA) had been listening to the PVO's air defense communications. In order to span the huge distances of the Soviet interior, and link the network with Moscow, the PVO was still using high-frequency (HF) radio. These HF signals could be intercepted by the NSA's listening stations ranged all along the Soviet southern border from Karamursel, near Istanbul, to Peshawar itself. Within these top-secret facilities, linguists who were fluent in Russian replayed the tapes, trying to follow and make sense of the major alert that had been called in the Soviet air defense system. Forty minutes after Powers was shot down, they intercepted the PVO report from Sverdlovsk which told Moscow headquarters that radar tracking of the unidentified target had been discontinued. Three hours later, in the small hours of Sunday morning in Washington, that news reached the CIA.[20]

Cover Story

Carmine Vito, the former Det A pilot who was the duty officer for Project CHALICE in DPD that Sunday, started working the phones. He rang Bob King. King notified Colonel Goodpaster and tried to reach Bissell, who was out of town for the weekend. High-level officials at the other government agencies who were cleared into the U-2 program had to be told: the State Department, the Pentagon, and NASA. Goodpaster called the President at Camp David and told him that the plane was missing. Bissell flew back to Washington, and went straight to Project HQ on H Street where he was met by a grim-faced Bill Burke. Hugh Cumming, who as head of the Bureau of Intelligence and Research was DPD's main contact at the State Department, was also there. They rehearsed the long-established cover story about a NASA weather reconnaissance mission having gone off course. Cumming requested that they delete all reference to the flight having started from Pakistan. The revised cover story had the plane straying off course—possibly—from a mission which was supposed to be conducted entirely within Turkish airspace.[21]

The next morning, Monday 2 May, Goodpaster took the cover story to Eisenhower for his approval. But the cover story—especially now that it had been amended—depended on the pilot being dead so that the fiction could not be contradicted. Allen Dulles and Dick Bissell had personally assured the President that a U-2 pilot would not survive a shoot-down. So the cover story was approved by the White House.[22]

In the countryside outside Sverdlovsk, the wreckage of the U-2 was slowly collected over the next few days. It was scattered over nine square miles, but the larger pieces were clearly recognizable as an aircraft, indeed as a spyplane. Film from both the tracker and the B-camera film was salvaged and sent for processing. Some of the SIGINT systems were virtually intact. A crane was summoned to recover the engine, which had fallen into a marsh. Some drunken Mayday celebrants from one village had taken their axes to one wing, which fell near them. The special

Lockheed test pilot Bob Schumacher pulls the oxygen line and faceplate for the MC-2 helmet towards him. Problems with the faceplate clasp and the oxygen system contributed to three fatal U-2 accidents in 1957-58, forcing a redesign of the pilot's vital life support system. (Lockheed C88-1447-76)

After Chris Walker was killed during U-2 pilot training in July 1958, Squadron Leader "Robbie" Robinson was sent to replace him as RAF detachment commander. Robinson was already familiar with high-altitude flying, having piloted the modified Canberra B.2 which could reach 70,000 feet, thanks to a Scorpion rocket booster. That aircraft forms the background to this posed photo, in which Robinson improbably wears a collar and tie together with his partial pressure suit! (Paul Lashmar collection)

The large airscoop for the F-2 sampling system is visible below the forward fuselage of this Detachment C U-2A landing at Atsugi.

The CIA's U-2 detachments relied upon extensive contractor support, especially by the Lockheed maintenance crews. Bob Murphy (Top left) was field service manager at both Dets C and B in the late fifties, supported by long-serving tech reps such as Jim Wood (Top Right). Kelly Johnson appointed Ed Martin (Bottom Left) as overall U-2 project manager in early 1958. He was succeeded by Art Bradley (Bottom Right) in the fall of 1959. (via Jim Wood and Ed Martin)

Behind the fiberglass nose panels, most early U-2s carried SIGINT systems 1 and 3. Article 368 was retained at Edwards North Base for various tests through 1957 and 1958, but was eventually assigned to the USAF as 56-6701. (Lockheed C88-1447-54)

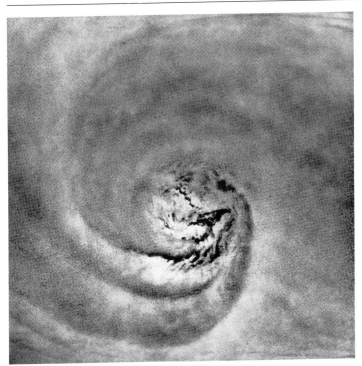

On 25 September 1958, Detachment C pilot Tom Crull flew a "genuine" weather reconnaissance mission. From 70,000 feet, he brought back a valuable series of photographs of Typhoon Ida, an intense storm which caused major damage in Japan. (Weatherwise)

The first aircraft to be modified with the Baird Atomic infrared sensor for detecting aircraft or missiles was 56-6722. It is seen here in late 1958, after the rotating barrel was moved from the lower to the upper equipment bay. (Lockheed Report SP-112)

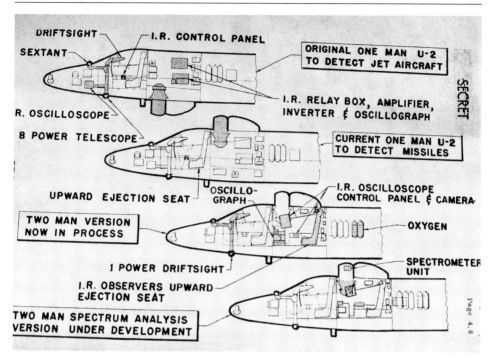

Evolution of the missile-sensing U-2 variants is shown in this drawing. Early flights of the one-man version showed that a second crewmember was needed to satisfactorily operate the sensor. (Lockheed Report SP-112)

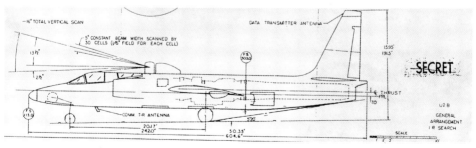

The U-2B would have provided a round-the-clock watch for Soviet offensive missile launches against the U.S. Note the conventional tricycle undercarriage. Kelly Johnson's proposal to build a patrol fleet of more than 80 U-2Bs was taken seriously, but ultimately dropped as early-warning radars were built, followed by missile-warning satellites. (Lockheed Report SP-112)

A U-2 "back-seater" prepares to enter the cramped confines of the modified payload bay, where the controls and displays for the infrared sensor were housed. Three of these two-seat aircraft were built in 1958-59, though the first soon crashed. The two survivors were eventually designated as U-2D models. (AFFTC History Office)

The Air Force Flight Test Center at Edwards formed a Special Projects Branch to fly the missile-detecting U-2s, and although the warning patrol fleet was never built, sensor development work continued into the 1960s. The early AFFTC pilots and observers are shown here, in front of two-seat U-2 56-6954. Majors Carpenter and Andonian (middle of front row) were the first two commanders. (AFFTC History Office)

Captain Bobbie Gardiner was one of four pilots from the 4080th SRW who flew peripheral photo reconnaissance missions along the USSR's Siberian borders in spring 1959. These long and difficult flights were codenamed CONGO MAIDEN, and were repeated the following year. (NARA 164462 AC)

This U-2 photograph shows the Potala Palace in Lhasa. The Tibetan crisis of March 1959 prompted a series of overflights by Detachment C, as the CIA stepped up aid for the Dalai Lama's supporters fighting the communists. (CIA)

The MiG-19PM (Farmer-B) *was thought to pose a greater threat to the U-2 than earlier variants. It carried an air-air radar which could guide four beam-riding K-5 (AA-1 Alkali) missiles. (Russian Aviation Research Trust)*

The runway at Incirlik AB, Turkey appears over the left wing slipper tank of a U-2, as an aircraft from the resident CIA Detachment B returns from a mission. Each slipper tank provided a precious extra 100 gallons of fuel. (via Stan Beerli)

Al Rand pulls 'mobile' duty at Incirlik, sometime in 1958-59. Each U-2 pilot took turns at this, which involved preflighting the aircraft, assisting the mission pilot into the cockpit, and then maintaining radio contact whenever the U-2 was taking-off or landing. (via Stan Beerli)

Last-minute adjustments to outer coveralls and pressure suit are made, before this U-2 pilot is helped into his aircraft. Once in the cockpit, his portable oxygen cylinder will be removed, as he is hooked up to the aircraft's supply. (Lockheed C88-1447-87)

Lockheed test pilot Ray Goudey gets his pulse checked by the flight surgeon, as he prepares for another high-altitude flight. Goudey made the first flight of the re-engined U-2C model, on 13 May 1959. (Lockheed C88-1447-73)

Compared with the J57, the Pratt & Whitney J75 engine offered a large increase in thrust, although it was 1,000 lbs heavier. Importantly, though, it could just be squeezed into the existing U-2 fuselage! (author)

This is the U-2C prototype (Article 342) standing on the dry lake near Edwards North Base in summer 1959, marked with its USAF identity 55675. Outwardly, there was no difference from the U-2A, except for the widened intakes needed to feed a greater air mass to the new engine's compressor. (Lockheed Report SP-179)

The SIGINT systems on the CIA's aircraft were upgraded in 1958-59. This U-2A is carrying the new System 3 and System 6 combination, which provided enhanced reception of COMINT and ELINT signals, while still allowing the primary camera payload to be carried. A new spiral C-band antenna is mounted on the forward equipment bay hatch, and the slab radome on the lower rear fuselage contains a scimitar antenna for VHF and A-band.

John Parangosky, the Agency's executive officer at Detachment B, flew to Iran in this C-47 to help secure landing rights for renewed U-2 overflights of the USSR (via Stan Beerli)

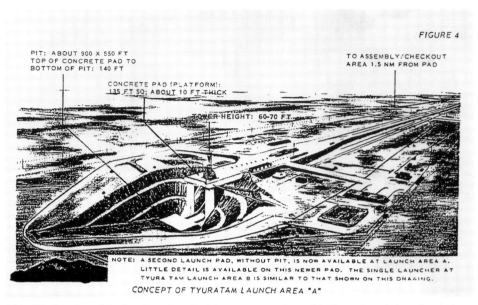

FIGURE 4

PIT: ABOUT 900 X 550 FT
TOP OF CONCRETE PAD TO
BOTTOM OF PIT: 140 FT

TO ASSEMBLY/CHECKOUT
AREA 1.5 NM FROM PAD

CONCRETE PAD (PLATFORM):
135 FT SQ; ABOUT 10 FT THICK

TOWER HEIGHT: 60-70 FT

NOTE: A SECOND LAUNCH PAD, WITHOUT PIT, IS NOW AVAILABLE AT LAUNCH AREA A.
LITTLE DETAIL IS AVAILABLE ON THIS NEWER PAD. THE SINGLE LAUNCHER AT
TYURA TAM LAUNCH AREA B IS SIMILAR TO THAT SHOWN ON THIS DRAWING.

CONCEPT OF TYURATAM LAUNCH AREA "A"

This CIA drawing of the first launch pad at Tyuratam, complete with dimensions, is based entirely on analysis of imagery returned by overflying U-2s. (CIA)

地に墜ちた
黒い天使

MISTERIOUS BLACK ANGELE

When Tom Crull crash-landed Detachment C's newly-acquired U-2C on a small airfield near Tokyo in September 1959, Japanese photographers were quickly on the scene. Despite the best efforts of project security officers, photographs depicting the "misterious black angele," which carried no identity markings, were soon published in Japan. (Aireview)

Operation Crowflight was reorganized in mid-1959, but the long, regularly-scheduled sampling flights continued. Officers and airmen assigned to the 4080th SRW's "home-based" detachment at Laughlin are shown here, complete with crow mascot. (via Tony Bevacqua)

After the British joined the CIA's Detachment B in Turkey in 1958, three training and demonstration deployments were made to the UK. Seen here in front of the supporting C-130 transport at RAF Watton in East Anglia are (from left) Maj Don Scherer (Det B operations officer), Sqn Ldr Robbie Robinson (RAF U-2 detachment commander), Col William Shelton (Det B commander); three members of the C-130 crew, and Dr John Clifford (the RAF flight surgeon at Det B). (Paul Lashmar Collection)

The Saratov airframe plant was a target on Sqn Ldr Robbie Robinson's overflight, 6 December 1959. (CIA)

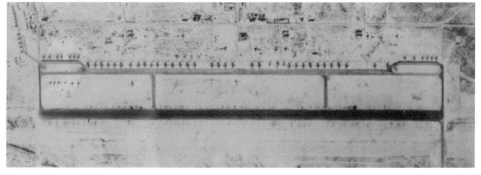

On the opposite bank of the Volga River to Saratov was Engels, and its bomber base. Mission 8005 captured a long line-up of Bison bombers on the snow-covered airfield. (CIA)

Left: The CIA's Jim Reber chaired the Ad-Hoc Requirements Committee (ARC), which nominated the targets for U-2 overflights. By early 1960, ARC had designated four areas of the northern USSR as the highest priority for coverage, since they might conceal operational Soviet ICBM bases. (CIA)

Below: This table from a Top Secret CIA report in 1960 illustrates the practical difficulty of searching for possible Soviet ICBM sites with the U-2. During the period specified, only four deep penetration missions were authorized and successfully conducted. Although each mission was about nine hours long, and the routes were planned to cover some of the "priority areas" thought most likely to contain ICBM sites, the actual percentage coverage from "useable TALENT photography" (i.e. cloud-free U-2 photography) was quite low.

Areas of the USSR Covered by Useable TALENT Photography
January 1959-June 1960

Area	Total Land Area (Square Miles)	Estimated TALENT Coverage (Square Miles)	(Percent)
Total USSR	8,647,000	650,000	7.5
Suitable for Deployment	4,764,000	650,000	13.6
Priority Areas Total	2,081,000	75,150	3.6
Area 1	467,800	10,750	2.3
Area 2	315,600	60,270	19.1
Area 3	170,700	4,130	2.4
Areas 4-8	1,126,900	0	0

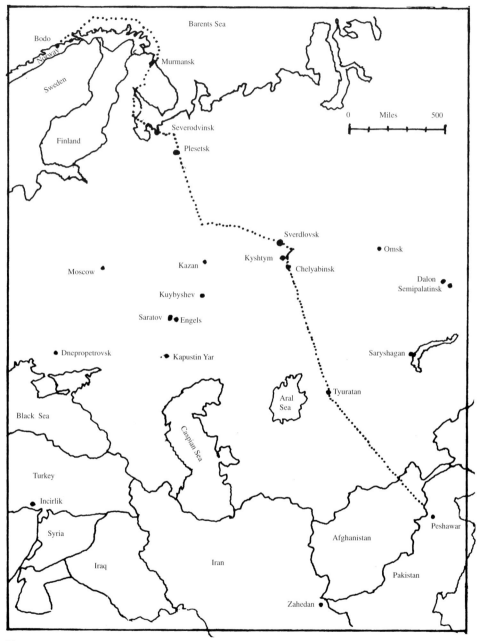

This map shows the key targets for the last five overflights. The Soviet capital, and the U-2 take-off and landing bases, are also shown. Dotted line is a simplified version of the flight track for Operation GRAND SLAM on 1 May—the first mission planned to fly all the way across the Soviet Union.

The V750 missile for the S-75 Dvina surface-to-air system was first shown in a Moscow parade on 7 November 1957. Western intelligence assigned the reporting name SA-2 Guideline. By 1960 the system had been deployed around most major cities and military-industrial complexes in the USSR.

An SA-2 missile battery had a distinctive "footprint" when photographed from above. Tracks linking the six missile launchers to the missile guidance radar and control van at the center of the site were laid out in a pattern which resembled the Star of David. (CIA)

56-6952 was one of three 4080th SRW U-2A models which flew the second series of peripheral photo reconnaissance flights codenamed CONGO MAIDEN in March-May 1960. On 1 May 1960, while the CIA's U-2 was flying over the Soviet heartland, the PVO was also trying to intercept one of these USAF aircraft, as it skirted the Kamchatka Peninsula.

The pilots of the last two U-2 overflights of the Soviet Union were Frank Powers (left) and Bob Ericson (right). Ericson was fortunate to return from his 9 April 1960 mission, after repeated Soviet attempts to shoot him down. On 1 May 1960, he strapped Powers into the cockpit for his ill-fated flight. (via Jim Wood)

At Project HQ in Washington, many of the key positions were filled by USAF officers on secondment. Colonel Stan Beerli (center) returned from Detachment B in August 1959 to head air operations. He brought with him Majors Ray Sterling (left) and Bill Seward (right), who both participated in planning for the last few U-2 overflights. Beerli flew to Bodo, Norway, in late April to head the recovery team for Operation GRAND SLAM. A wasted journey, as it transpired! (via Stan Beerli)

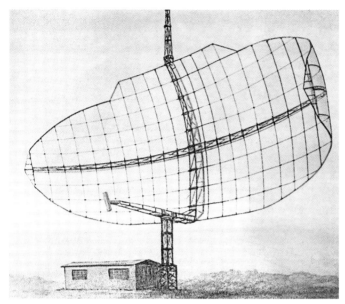

The extent of Soviet early-warning radar coverage was a key constraint when the last overflights were planned. The CIA hoped that gaps still existed along the southern border, allowing the U-2 to penetrate without detection. But powerful new search radars like this P-14 (Tall King) were now being deployed.

Detachment C pilot Bill McMurry force-landed his U-2 in a Thai rice paddy on 5 April 1960, after running out of fuel. Three weeks earlier, a USAF U-2 was dead-sticked onto a frozen lake in Canada. Fortunately, there had not been a serious technical failure over "denied territory"—yet. (CIA)

This was the communications station on its mobile trailer that Detachment B used to receive coded operational messages when deployed. At Peshawar on 1 May 1960, the "commo" technicians couldn't pick up the "go-code" on the pre-assigned frequency. It was eventually received "in the clear"—a theoretical breach of security. (via Jim Wood)

This U-2 is in the same configuration as Article 360 when it was shot down. There were no identifying markings on the almost-black paint. The slipper tanks provided precious extra fuel for the nine-hour mission. The B-camera in the equipment bay was the primary payload, augmented by SIGINT Systems 3 and 6.

The Sukhoi Su-9 interceptor was the PVO's best chance of intercepting the U-2 in the air. Unknown to Western intelligence, it had been rushed into service by spring 1960, and could reach 65,000 feet. But the aircraft sent to challenge Powers had no missiles, and was ordered to ram the target. (via Yefim Gordon)

Major Mikhail Voronov was in charge of the SA-2 battery which shot down the U-2. But his failure to report the successful action in timely fashion led to confusion amongst PVO commanders, and the order to fire more missiles.

Senior Lt Sergei Safronov was killed when his MiG-19 interceptor was hit by one of the PVO's own SA-2 missiles.

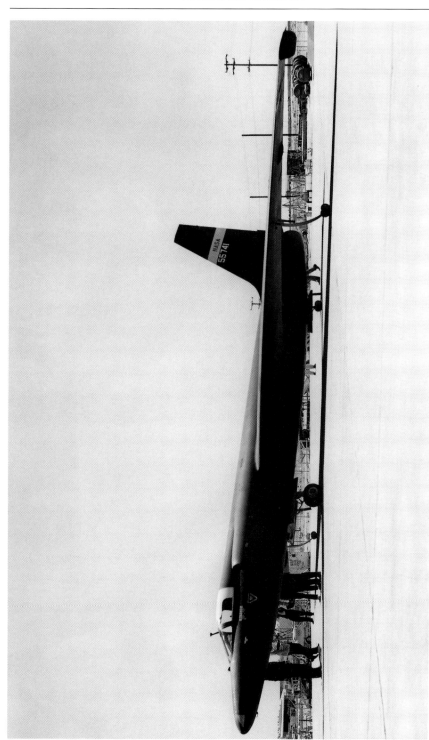

To satisfy the curiosity of American media in the wake of Powers' shootdown, this CIA U-2A (Article 378) was produced for their inspection at Edwards AFB. Since the cover story of weather research had not yet been discredited, a NASA tail band and false serial number had been hastily applied.

Wreckage from the downed U-2 was displayed in Moscow 11 days after the shootdown. All of the basic structure was recovered. (via Jay Miller)

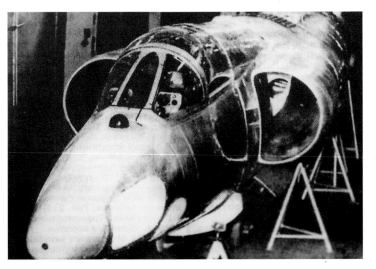

Looking much like the real thing, this mock-up of the U-2 was constructed by the Beriev Design Bureau. But Soviet attempts to fully replicate and fly a "U-2ski" designated S-13 were abandoned in 1961. (Russian Aviation Research Trust)

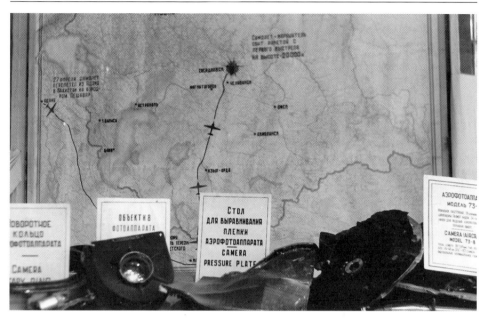

The Moscow display of U-2 wreckage also included the camera and SIGINT sensor systems. The spool for the 70mm tracker camera and various parts of the B-camera were placed on a table, in front of a map of the doomed flight.

Just two months after the U-2 was shot down, one of SAC's RB-47H aircraft flying a SIGINT mission close to the Kola peninsula was downed by Soviet fighters. It was a further embarrassment for the Eisenhower administration, which subsequently moved to exert tighter control of military peripheral reconnaissance missions like this and the CONGO MAIDEN series. (NARA 178161 USAF)

The super-secret National Security Agency (NSA) intercepted Soviet air defense communications from sites along the border, such as this USAF-run facility at Samsun, Turkey. But NSA linguists jumped to the wrong conclusions when they analyzed fragments of the PVO's confused chatter during the Powers shootdown. (USAF Air Intelligence Agency)

In March 1962, a month after his release from a Soviet prison, Frank Powers testified before the Senate Armed Services Committee. Days earlier, a CIA Board of Enquiry had cleared him of all blame for the MayDay incident, and praised his conduct during detention.

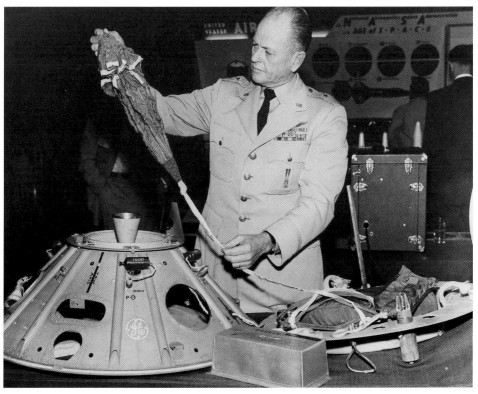

After many failures, the CORONA reconnaissance satellite system finally returned a film capsule from orbit on 19 August 1960. A recovery package is here being examined by Brig Gen Ozzie Ritland, who, six years earlier, had been senior USAF officer assigned to the nascent U-2 program. (AFFTC History Office)

FIGURE 3

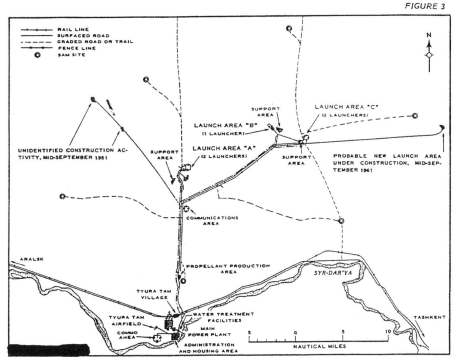

TYURATAM MISSILE TEST CENTER (Status in late 1960-early 1961)

Slowly but surely, the superior area coverage of the CORONA satellite imagery began to disprove the missile gap. Only three new Soviet ICBM sites were identified by mid-1961, and these were all still under construction. Good coverage of Tyuratam was obtained in December 1960 and September 1961, when this map was issued by CIA. It shows the rail spur leading north to the original R-7 launch pads, and the new launch areas served only by road, which had been discovered by the last successful U-2 flight over the USSR on 9 April 1960. Note the placing of no fewer than six SA-2 sites around the missile complex by this time!

Opposite: *Far from being consigned to history by the MayDay incident, the U-2 proved invaluable over China, Cuba, Vietnam, and elsewhere. In 1967, an improved and enlarged version designated U-2R (foreground) was flown. The CIA and USAF each ordered six, and when the Agency's U-2 operation finally closed in 1974, the 11 survivors all continued in military service. They were joined by a further 35 aircraft between 1981 and 1989, when the USAF ordered the U-2 into production for an unprecedented third time. (Lockheed U2-91-014-8)*

Today, the 99th Reconnaissance Squadron of the USAF's 9th Reconnaissance Wing continues the U-2 tradition, operating detachments around the globe. The 37 surviving U-2s (including two operated by NASA) have all been re-engined with the modern GE F118 turbofan, and seem set to fly on for many years to come. (Lockheed Martin)

fuel spilled out, thus making it more difficult for Soviet military experts to analyze it.[23]

PVO Shortcomings Suppressed

The PVO set up a commission to investigate the shootdown. But Marshall Biryuzov realized that this was no time to advertise the PVO's shortcomings. These included a severe shortage of trained manpower and missiles, and the lack of co-ordination—even jealousy—between the PVO's fighter interceptor troops led by General Savitsky, and the missile troops led by General Kuleshov. This went all the way down the command, and was a major factor in the confusion over Sverdlovsk on 1 May. The commission decided that in the hurry to dispatch the Su-9 and two MiG-19 interceptors, no-one remembered to change their transponder codes. These were still set to the mode for April, but it was now the first of May. As a result, Shugayev's missile battery didn't receive the proper IFF response when it interrogated the fighters, the commission concluded. That helped Shugayev conclude that he was tracking the intruder at 36,000 feet, rather than the PVO's own interceptors.[24]

It seems that Biryuzov managed to ensure that Premier Khrushchev never heard about the multiple missile firings, and the shooting-down of the MiG.[25] But word soon spread within military circles in Moscow that the official version of events was incomplete. Officially, Powers had been shot down by a single missile above 20,000 km (65,000 feet). That was true, of course, but it wasn't the whole story. According to some rumors, the PVO fired as many as 14 missiles over Sverdlovsk that day. The full story related in these pages emerged slowly from 1990, as various participants recounted their part in the action to bring down the U-2.[26]

On Tuesday 3 May, NASA released the CIA's cover story. It was a detailed fiction, describing how a NASA U-2 research plane "apparently went down" after the pilot "reported he was experiencing oxygen difficulties." A search was being conducted in the area of Lake Van, eastern Turkey. To maintain the fiction, the C-54 transport used by Det B took off from Incirlik, with orders to search the area in question.

Unfortunately, no-one had informed Powers before he took off that oxygen difficulties formed part of the cover story. In any case, the U-2 pilot had soon abandoned the fiction of a weather reconnaissance flight which had gone astray. The weight of evidence to the contrary was overwhelming. While he was still in Sverdlovsk, his interrogators had recovered incriminating evidence from the wreckage and shown it to him: his flight maps over the USSR, his survival pack, which included 7,500 Russian roubles, even the tracker camera with its exposed film. Powers decided that he had little option but to tell the truth to his captors' questions, with some limitations. But he would not volunteer any information they did not ask for.[27]

Soviet Strategy

In Moscow throughout Wednesday, the Central Committee of the Communist Party debated how to handle the incident. Khrushchev proposed that he reveal only that the plane had been shot down, *not* where, *nor* that the pilot had been captured. He planned to make political capital out of the incident, by forcing the U.S. to elaborate on the cover story, before blowing it apart. It was news management on a grand scale, and it worked. At a scheduled meeting of the Supreme Soviet the next day, Khrushchev revealed that an American spyplane had been shot down and asked rhetorically who had sent it. In Washington, the press demanded a response. In a second statement, NASA added some detail to the cover story. The State Department told reporters that the pilot might have lost consciousness over Turkey, allowing the plane to "continue on automatic pilot for a considerable distance and accidentally violate Soviet airspace."

At a Moscow reception on Thursday evening attended by the U.S. Ambassador, a Soviet diplomat let slip that his government was "still questioning the pilot" who had been shot down. This was the first indication that Powers was still alive. The bad news was immediately cabled to Washington. On Friday 6 May, the State Department categorically denied that there had been a deliberate violation of Soviet airspace. DPD scrambled to concoct an alternative cover story. Dick Newton, the executive officer at Det B, was nominated as the fall guy. He would confess to exceeding his authority by sending Powers into Soviet airspace. The CIA official was hurriedly flown from Turkey to Germany and hidden in an Agency safe house, where he couldn't be reached by reporters.[28] Meanwhile, a Moscow newspaper published a photo showing a heap of twisted wreckage, which was identified as "the pirate plane."

The American newspapers had been bawling for a close look at the U-2. On Friday their wish was granted, when a photo call was arranged at Edwards. Not at the top-secret North Base site, though. The gentlemen of the press were taken to a ramp on the main part of the flight test base. The CIA unit towed one of their U-2A models (Article 378) over from North Base. It had been hurriedly painted with a yellow NASA tail band and a fictitious serial number. The photographers clicked away, but the groundcrew remained tight-lipped when quizzed about the plane.

In fact, no-one cleared into the project at Edwards was allowed to discuss the shoot-down, even amongst themselves within the secure confines of North Base. Project HQ told unit commander Lt Col "Rosy" Rosenfield to clamp all mouths shut. "What was being said on the morning talk shows, became classified information the minute it was discussed on the base! We pilots were the only ones in the whole of the Antelope valley that didn't have anything to say about the incident," noted one frustrated U-2 pilot.

The Awful Truth

On Saturday 7 May, the Soviet Premier returned to the Supreme Soviet, to wind up its meeting. He now revealed the awful truth: "We have the remnants of the plane—and we also have the pilot, who is quite alive and kicking!" Khrushchev held up a photograph of an airfield which he claimed was taken by the U-2's camera. Again, he speculated that "American militarists" had ordered the flight, rather than President Eisenhower himself.[29]

Of course, the Soviet Premier caused an immediate sensation. Headlines screamed from the Sunday newspapers around the world. In Washington, Allen Dulles offered to resign and take the rap. On Secretary of State Herter's advice, a statement was issued that admitted the U-2's mission as a border surveillance tool. The aircraft had "probably" flown into Soviet airspace, but this had not been authorized in Washington. The statement did not play well with the American press, let alone in Moscow. Was the White House in charge here, or not?

Finally, at Eisenhower's insistence, the State Department admitted on Monday 9 May that the President had authorized the entire U-2 program as a means of gaining adequate knowledge of the Soviet military-industrial complex. But specific missions had not been subject to the President's authorization, the statement lied. There were more sensational headlines around the world.

At Eielson airbase in Alaska, SAC U-2 pilot Dick Leavitt heard about the shootdown of Powers with even more interest than the rest of the 4080th wing contingent deployed there. A few days earlier, Leavitt had flown the 11th and last CONGO MAIDEN flight. He had paralleled the Soviet coastline, photographing airfields and radars all the way to Petropavlovsk on the Kamchatka Peninsula. Soviet fighters chased him for several hundred miles. He had taken off from Eielson around 11 am on 30 April, and returned at 8 pm—another very long mission.[30] But he had crossed the International Date Line, so it was Sunday 1 May in Siberia when the flight took place. Leavitt realized that he had been airborne at the very same time as the CIA's U-2. Powers was not the only U-2 pilot to have stirred up the Soviet air defenses on their national holiday!

In Washington Art Lundahl—the master briefer—pulled off one of his best performances as 18 leading Congressmen were privately told the extent of the U-2 program and its accomplishments. In Burbank, Kelly Johnson told reporters that the photo of the wreckage which had been released in Moscow was a fake. But Johnson unaccountably went further. "I do not believe they shot down the U-2 by either a missile or another aircraft. If they have the U-2, it is because some mechanical or oxygen failure caused it to descend far below its normal cruising altitude."

In Moscow that same day, Monday 9 May, newspapers published a decree of Supreme Soviet. Honors were awarded to 21 PVO officers who had been involved in the shootdown. Voronov, Sheludko, and Safronov headed the list: they were given the Order of the Red Banner. The fact that the MiG-19 pilot's award was posthumous was carefully omitted from the official record.

Propaganda

The wreckage of the U-2 was brought to Moscow. On Wednesday 11 May it was put on display in Gorkiy Park. Three months earlier, President Eisenhower had foretold exactly this scenario, if a U-2 were to be shot down. Trailed by reporters, Premier Khrushchev was one of the first visitors to the unusual exhibition. He was milking the U-2 incident for all it was worth. The U.S. ambassador in Moscow concluded that the Soviet leader would turn the forthcoming Paris summit into a propaganda event. President Eisenhower faced the press in the White House and talked of "the distasteful but vital necessity" of intelligence gathering.[31]

The Soviets also turned the heat up, on those other countries which were clearly associated with the U-2 program. Because of their direct involvement with the 1 May mission, Norway, Pakistan, and Turkey were all in the firing line. The State Department spokesman claimed that those countries knew nothing about U-2 missions into the USSR. "These planes come and visit our country. How do we know where they go after they leave?" declared Ayub Khan, the Pakistani leader (who knew exactly where the spyplane was going). The Shah of Iran denied all knowledge of the U-2 (and the use of the Zahedan airbase to recover two previous overflight missions did remain a secret).[32]

The Norwegian foreign minister cross-examined the intelligence chief Wilhelm Evang, and the Bodo-based Norwegian air force commander Tufte Johnson. Both denied that they knew anything about a plan to overfly the USSR with the U-2. Evang was furious with the Americans; the CIA station chief in Oslo had deliberately disappeared, and the Norwegian had to rely on the newspapers to discover what the U.S. was saying about the shootdown, and Norway's involvement. Oslo sent a secret protest note to Washington. The government issued a statement, in which Tufte-Johnsen was quoted saying that landing permission for "that airplane" at Bodo had neither been requested nor granted.[33]

In London, Secretary of State for Air George Ward summoned the RAF U-2 pilots to his office. The British fliers had been quickly evacuated from Turkey as soon as news of the shootdown reached London. But the British press was picking up rumors that the RAF was somehow involved in the U-2 overflights. Opposition Members of Parliament (MPs) had tabled some questions in Parliament. The Air Minister wanted to know if Powers would spill the beans to his Soviet captors about the British group. The Minister told the pilots that he was torn between lying to Parliament, or saying nothing. British U-2 detachment commander Robbie Robinson described Powers as a pleasant, quiet, and likable American. He told Ward that Powers probably would reveal the RAF involvement, but he couldn't be sure. Ward told the British fliers to disappear from view for a few months, just in case.[34] In Parliament on 11 May, Ward refused to answer when an MP asked him to confirm the "co-operation between the RAF and the USAF for obtaining photographic information about Russian air and military bases."[35]

In Japan, the U-2 incident threatened to scuttle a new security treaty with the U.S. The treaty was being discussed by the Diet, the Japanese parliament, amidst a growing uproar over the presence of the "mysterious black-painted spyplanes" just outside Tokyo. "There is no truth to reports that a U-2 aircraft conducted intelligence missions from Japan," said the U.S. State Department. Again, this wasn't true.

SAC's 4080th wing tried to continue as normal. The next phase of the atmospheric sampling program was getting underway, and it was a complicated exercise. For the first time, U-2 sorties from three separate bases (Ezeiza, Laughlin, and Eielson) would be co-ordinated. Martin JB-57 Canberras would also be sampling at lower levels at the same time. Three U-2As left Laughlin for Ezeiza, Argentina, on 3 May, but were held at Ramey AFB while higher authority debated whether they should continue. Meanwhile, the CIA wanted to get its Edwards-based U-2 pilots out of the way. They were dispatched to Ramey and flew the SAC aircraft for a month on the co-ordinated sampling flights. "Talk about boring: man, those flights were the living end!" said one of the CIA pilots.

Interrogation

It certainly wasn't as exciting as being shot down over denied territory! In Moscow, Frank Powers was undergoing intensive daily interrogation. But he was not physically abused, and as the questioning continued, Powers realized that he could successfully keep some important details of the U-2 program secret. His task was complicated by the revelations about the program which were now occurring back home. The interrogators could trip him up, when he denied something that was then admitted in Washington. Incarcerated in the Lubyanka prison, it wasn't easy to second-guess!

Nevertheless, Powers did manage to withhold significant information. The Soviets never asked about any British pilots, so he didn't tell them. They accepted that he had not flown over the USSR before, when he once had. They didn't press him on flights over other countries, such as the wide-ranging Middle East missions that Det B regularly flew. He successfully convinced them that he knew little about the cameras, and nothing about the SIGINT systems. In general, Powers played the role of the poor dumb pilot. His superiors in the detachment only told him what he needed to know for the flight, he explained to the Soviets.

(To some extent, this was true. Officially, only the pilot actually assigned to each overflight mission was briefed. The CIA's security officers encouraged a tacit understanding amongst the pilots that, by and large, they wouldn't enquire of each other's missions in great depth. That way, if one of them did eventually fall prey to Soviet defenses, they would have less to tell. Moreover, the nature of the targets to be overflown were not explained to the pilots in great detail. Nor did they get much feedback on the intelligence results from their flights).

Powers was asked to reveal how high the U-2 could fly. The pilot thought of his colleagues who might, even now, be sent on similar missions. He wanted to deny the Soviet air defenses, the possibility of setting their missiles to detonate at the correct altitude. So Powers said that he had been shot down at 68,000 feet, instead of 70,000 feet—a small lie. He claimed that this was the plane's maximum altitude—a bigger lie. The interrogators didn't challenge him. After all, the PVO radar tracking was so confused, they had no serious evidence to the contrary.

In Washington, though, the intelligence community jumped to their own conclusions. The NSA's intercepts of the Soviet air defense reporting network were played and replayed. Some of the PVO reporting related the mistaken notion which originated in Shugayev's battery, that the U-2 had descended to 36,000 feet. The State Department weighed in with a report from the U.S. embassy in Moscow, on the exhibition of wreckage in Gorkiy Park. "The unusually good condition of debris from the crash, and its reported disposition over a nine-mile-wide area, was inconsistent with it having fallen from 68,000 feet," the report said.

Summit Failure

The summit meeting in Paris in mid-May was a disaster. Khrushchev demanded that Eisenhower denounce the U-2 flights over the USSR as provocative. Ike had said as much to Dulles, Bissell, and the others in private, but he refused to apologize for them in public. He did tell the Soviet Premier that there would be no more flights during his administration, which still had nine months to run. This wasn't good enough for Khrushchev. He withdrew an invitation for Eisenhower to visit the USSR, and walked out. Art Lundahl and Jim Cunningham played a bit part in the Paris summit. They were summoned from Washington when Eisenhower decided to show General De Gaulle some U-2 imagery. At a briefing in the Elysee Palace, the French leader showed considerable interest in the photographs of Soviet nuclear and missile installations.[36] (He probably didn't know that the U-2 had recently been spying on France's own nuclear weapons program. An aircraft from Det B had flown over the test site at Reggane in Algeria where, on 13 February 1960, the first French nuclear test took place).

Back in Washington on 24 May, the President called a meeting of the National Security Council (NSC) and reviewed the whole sorry affair. He observed that the previous overflights had been so successful that "we may have become careless."

He previewed what information should be made available to the Senate enquiry which was pending. He didn't want it made public how many overflights there had been, nor that he had approved specific missions. Most of all, he didn't want anyone to admit that other countries had been a party to reconnaissance overflights.[37]

Eisenhower insisted that he had previously been briefed to the effect that, "the pilots on such flights were taught to destroy the plane rather than let it fall into enemy hands." He would never have approved the cover story, if he hadn't believed this.

Apparently, Powers had not activated the destructor. Why not? It was a legitimate question, but the President had been sorely misled by whoever told him that the plane would be destroyed. The destructor was a 2.5 lb. charge of cyclonite which was placed in the equipment bay behind the cockpit. Its purpose was to destroy the camera—the most incriminating evidence. There was some hope that the explosion might lead to further break-up of the wreckage inflight, but this theory had never been tested. However, it is doubtful whether a 2.5 lb charge could have completely destroyed even the camera, let alone its huge, tightly-wound film magazines.

The control panel for the destructor was mounted on the forward right canopy rail; the pilot threw one switch to activate the system, and a second switch to start the timer, which was usually set for a 60-second delay.[38] The delay was provided so that the pilot could have enough time to escape after setting the timer, since he might otherwise be injured in the blast. Powers would later say that he had considered setting the timer before he tried to escape from the disabled U-2. But then he decided to ensure that he really could get out of the cockpit first. Unfortunately, when he released the seat belt, g-forces threw him half out of the cockpit and he was unable to reach the switches.

The NSC meeting continued with more ill-tempered remarks and unfounded speculation from the nation's leadership. "Apparently Powers started talking as soon as he hit the ground," the President remarked irritably. Then DCI Allen Dulles chimed in, noting that "we traced the U-2 piloted by Powers down to 30,000 feet." This was from the NSA intercepts, of course. Bullets had been fired at the plane while it was in the air—the pictures of the wreckage in the Moscow exhibition showed this, Dulles asserted.

Eisenhower and Secretary of State Herter told the meeting that they were worried about revealing the "fact" that the U.S. had tracked the U-2. This was a highly secret SIGINT capability—officially the National Security Agency didn't even exist! In reality, though, the NSA hadn't "tracked" anything. As we have seen, it had intercepted fragments of the PVO's long-range communications. Herter's deputy, Douglas Dillon, recommended that they use the wreckage photos to make the case that the U-2 had descended before being shot down. But, surprisingly enough, the DCI was more willing to reveal the SIGINT capability. Dulles explained that the

USSR was claiming that their rockets could shoot a plane down from 60-70,000 feet. It would be re-assuring to our allies if we could inform them that the plane had *not* been shot down at this high altitude, he continued.[39]

There was the rub! It was not politically expedient in Washington to admit that the Soviets had a high-altitude SAM that worked. For a start, SAC's subsonic nuclear bombers could be sitting ducks if they attacked the USSR. How could America reassure its allies on the frontline of the Cold War that the nuclear deterrent would hold Soviet aggression in check? On Capitol Hill, moreover, questions might be asked about all the money being spent on the supersonic XB-70. Could it, too, be vulnerable?

One day later, Eisenhower made a television address explaining the summit failure. He displayed a B-camera photograph of the NAS North Island, San Diego. "It was taken from more than 70,000 feet...the white lines on the parking strips around the field are clearly visible," he told viewers.

Flameout Story Grows

The next day, 26 May, Congressional leaders took breakfast in the White House. The President told them that he didn't believe the Soviets had shot down the U-2 with a rocket. "The plane's engine had flamed out," he believed. It was "obvious" that the bullet holes shown in the photographs "must have been put in the wing at a lower altitude," possibly by Soviet fighters.[40]

On 27 May, the Senate Committee on Foreign Relations began four days of closed-door hearings on the U-2 affair. Herter testified on the first day; among other comments, he confirmed that the U.S. Ambassadors in Turkey and Pakistan knew nothing of the flights (all the liaison went through the CIA station chief's office). On 31 May, a delegation from the CIA briefed the Senators. DCI Dulles did most of the talking, though his deputy, Pierre Cabell, was also there. So were Art Lundahl and Dick Helms, Bissell's deputy at DDP.

Dulles told the senators that the main intelligence targets for the U-2 flights had been the Soviet bomber force, air defense system, and its missile, atomic energy, and submarine programs. He showed imagery of Kapustin Yar and Tyuratam, noting that the U.S. had learned much about the Soviet doctrine of ICBM deployment, as well as technical characteristics of the missiles themselves. In the atomic energy field, the U-2 coverage had included the production of fissionable materials, weapons development and test activities, and the location, type, and size of many Soviet nuclear stockpiles. The photography had provided the first firm information on the magnitude and location of Soviet domestic uranium ore mining and processing activities, so that its production of fissionable material could be estimated.

The material obtained had been used to correct military maps and aeronautical charts, Dulles continued. Hard information about the nature, extent, and location of Soviet ground-to-air missile development had been obtained. Much had been learned about Soviet air defenses in general, and early warning radar development. Scores of airfields had been photographed. U-2 photography provided new and accurate information to SAC's strike bomber crews, Dulles noted, making it easier for them to identify their targets and plan their navigation. "In the opinion of our military, our scientists, and senior officials responsible for our national security, the results of this program have been invaluable," Dulles concluded.

The questions began. When the sensitive issue of whether the President had personally approved each overflight was raised by the senators, Dulles hedged and dissembled. But he had no such inhibitions about describing the shootdown. The U-2 "was initially forced down to a much lower altitude by some as-yet undetermined mechanical malfunction," he said. "If the plane had been hit by a ground-to-air missile, it would have disintegrated," he added.[41] It seems that no-one at CIA headquarters could believe that a SAM might disable a U-2, rather than outright destroy it. Towards the end of his testimony, Dulles did concede that a Soviet SAM "might have had a near miss...(which) caused (the U-2) to lose altitude." But no-one seemed to take this theory seriously.

Concerning the destructor, Dulles did make clear to the senators that "no massive device capable of ensuring complete destruction could be carried in this aircraft." He also confirmed that pilots "were not given positive instructions" to take their own lives with the poison pin. Suggestions had already been aired in public that Powers should have jabbed himself with the poison pin to avoid capture, like any self-respecting spy. That was never the intent of providing the pilots with the pin; it was provided only as a means of suicide in the face of unbearable torture.

Dulles therefore laid to rest, for the Senators' benefit, a couple of U-2 myths-in-the-making. But this was a closed session of the committee. Only a heavily-censored transcript of the proceedings was issued. The media continued to peddle rumor and misinformation about destructors and poison pins as fact. Moreover, Dulles' inaccurate testimony to the Senators about the circumstances of the actual shootdown soon leaked. Within a week, headlines read "Soviet Radar Defense System Far From Perfect, Ineffective at Altitudes at Which U-2 Flies" and "Despite Boasts, Plane Could Not Have Been Hit at 65,000 feet."[42]

The story grew with the telling. Soon, Powers was said to have had a flameout, and to have made three or four attempts to restart, none of which worked. One version even offered direct quotes from Soviet radar operators as proof: "He's coming down, he's still descending!" Another version held that Powers himself had radioed back to base that he was in trouble. This ignored the fact that the CIA's U-2s were equipped only with a short-range UHF radio, transmissions from which could not possibly be picked up from so deep inside the USSR.

Kelly Johnson wanted to test the flameout theory. At North Base, Lockheed test pilots made four flights in the prototype U-2C to check the restart characteristics of the J75 engine. Skunk Works engineers tested the destructor, analyzed photos of the ejection seat from Moscow, and reviewed the data from previous U-2 accidents. The wreckage as displayed in Moscow showed some similarity with that from a USAF U-2 which had broken up at about 35,000 feet. On 24 June, Johnson concluded that Powers "did not have enough time to (activate) the destructor switch or even use the ejection seat." This was partially correct. He further concluded that "the airplane was hit at less than normal altitude by a rocket." This was only correct in the sense that during their descent to the ground, some parts of Powers' plane may have been hit by the second, third, and fourth missiles fired by the PVO. Johnson still discounted the possibility that the aircraft had been disabled by a missile at cruising altitude, even though the technical report by his own engineers on the four new test flights was careful to note that "there is no direct evidence at this time that a high-altitude engine blow-out occurred." (The technical report was signed off by Art Bradley, who had taken over from Ed Martin as U-2 program manager for Lockheed nine months earlier).[43]

No More Overflights

Ed Martin was one of a growing band of ADP engineers whom Kelly Johnson had transferred to the A-12 project, now that the supersonic successor to the U-2 was firmly under contract. But at the highest level, even this project was threatened by the fallout from the U-2 incident. In early June, the President told Colonel Goodpaster that "he did not think the project should now be pushed at top priority." After all, he had already told Khrushchev that there would be no more overflights, at least not on his watch. Maybe the Mach 3 spyplane could go forward "on low priority, as a high performance reconnaissance plane for the Air Force in time of war," he mused.[44]

Now that the U-2 was out in the open, Kelly Johnson didn't see any point in keeping scores of workers out at the isolated North Base site. In mid-June, he moved the IRAN (Inspection and Repair as Necessary) program for the USAF's U-2 fleet to Burbank.

On 1 July 1960, one of SAC's RB-47H reconnaissance aircraft was shot down by Soviet fighters in the Barents Sea while flying parallel to the Kola peninsula on a SIGINT mission. The co-pilot and navigator were picked up from the water by the Soviets. The four other crew members perished. Premier Khrushchev claimed another "gross violation" of Soviet airspace, but the NSA knew otherwise, from SIGINT intercepts. The U.S. refuted the charge at the United Nations with a detailed map of the aircraft's offshore route.

The RB-47H had taken off from Brize Norton airbase in the UK. The new incident prompted a fresh round of Parliamentary questions to British ministers about the Anglo-American security relationship in general, and the U-2 in particular. Despite his reputation for unflappability, Prime Minister Harold MacMillan was now quite worried that British participation in the U-2 program would come to light. The USSR had announced its intention to put pilot Powers on trial in Moscow for espionage. Would the UK's biggest intelligence secret be revealed there?

British Minister's Conundrum

Already, left-wing MPs had discovered that Squadron Leader Chris Walker was killed during U-2 training at Laughlin two years earlier. The Air Minister George Ward told Parliament that "it is quite customary for RAF aircrew...to fly USAF aircraft, particularly advanced types." In private, Ward was still wrestling with his conscience. "To give information to the House would also...give it to our potential enemies. As a Service Minister...this places me in an awkward dilemma, from which I could only escape at the expense of betraying...the national interest," he wrote.[45]

Still, the British government stuck to the party line. When MacMillan's turn to face MPs came on 12 July, he retreated to the time-honored formula. "It has never been the practice to discuss (intelligence) matters in the House, and I have come to the conclusion that it would be contrary to the public interest to depart from precedent on this occasion."[46] President Eisenhower wrote privately to congratulate MacMillan: "I thought you handled this matter very well indeed. It is clear that we are up against a ruthless Soviet campaign against our free world bases."[47] Ike admired the British ability to keep their mouths shut, and told colleagues the U.S. should have done likewise after 1 May.

On Sunday 11 July, 10,000 Japanese demonstrators gathered outside NAS Atsugi, demanding that the U.S. remove the U-2s stationed there. In fact, the Japanese government had already requested their removal. The two aircraft were hastily dismantled and flown out by transport planes. There wasn't even time to send for the proper wing and fuselage transport carts: the Lockheed maintenance crews grabbed some old mattresses to use as protective padding for the disassembled parts. It was an ignominious departure, but the U-2 incident had created an uproar in Japan. By now, the three remaining aircraft in Turkey had also been hauled back to the U.S. by C-124s. Project CHALICE was facing an uncertain future.

The USAF weighed in with some blatant Monday-morning quarterbacking. General Nathan Twining, former USAF CinC and now Chairman of the JCS, asserted that the CIA "got too big for their britches. They did not know how to handle this type of operation...they made a mistake and went in twice from the south. We screamed and yelled, and we insisted that they change the plan and come in over Norway and Sweden and get closer to the target area." Twining went on to complain that Eisenhower should not have approved the CIA's running the U-2 program in the first place. Bissell was "a smart young man who wanted to take over, and he did."[48]

Twining conveniently ignored the fact that blue-suiters made up a significant proportion of the AQUATONE and CHALICE operation. Moreover, by formal agreement Bissell's deputy at DPD was appointed by the USAF. Indeed, USAF vice-commander and former SAC CinC General Curt LeMay had personally selected Colonel Bill Burke to take the job in 1958. He wanted one of "his" people to be at the helm in Washington, to exert as much military influence as possible.[49]

All this time, the U.S. Embassy in Moscow had been requesting access to Powers. He was kept in solitary confinement in the Lubyanka Prison. Except for a few heavily-censored letters, the U-2 pilot had been denied all contact with the outside world. The trial was set for 17 August, and a Soviet lawyer was appointed to defend him. "He specialized in losing state cases," Powers noted dryly.[50] The USSR continued to deny access to the pilot.

Inside Information

Late on the evening of 12 August 1960, two young American tourists were approaching the bridge over the Moscow River, just south of Red Square. A Soviet citizen approached them, began a conversation, and walked on with them. He was Oleg Penkovsky, the Colonel in Soviet military intelligence who was to become the West's most valuable spy in the Soviet military heirarchy. This was his first attempt to make contact with Western intelligence. During his conversation with the students, Penkovsky told them that he knew exactly what had happened to Powers and the U-2. The downed pilot was due to go on trial in Moscow just four days later.

Penkovsky had been the GRU duty officer on 1 May, when Powers was brought to Moscow. But the KGB had kept Soviet military intelligence at arm's length during the U-2 pilot's early interrogations. To the American students, Penkovsky repeated the details that he had learned second-hand from various GRU, KGB, and PVO sources. He confirmed that Powers had been shot down at high altitude, after his aircraft was disabled by a near-miss from a Soviet surface-to-air missile. Penkovsky related the story that a total of 14 missiles had been fired. He revealed that one of the missiles had inadvertently shot down a MiG-19 interceptor.[51]

Show Trial

The show trial of Francis Gary Powers was held in Moscow's impressive Hall of Columns from 17-19 August. The indictment was read out—a litany of propaganda complaints interspersed with occasional "facts." Curiously, the indictment made repeated references to the U-2 having "remained at an altitude of 20,000 meters (65,000 feet) throughout the entire flight".[52] This was even lower than Powers had admitted, but probably more politically acceptable to the Soviet fighter lobby! The fact that a high-altitude aircraft could burn off fuel during a long flight and thus rise higher in a "cruise-climb" was never mentioned, though it must surely have been obvious to Soviet aviation experts.

During cross-examination, Powers explained that point. He also repeated for the benefit of U.S. government representatives in court, that he had been shot down at 68,000 feet, which was the "maximum altitude" for the plane. Most of the information about Project CHALICE that Powers had revealed during his interrogation was repeated in court. It should have been obvious to knowledgeable observers that he had kept plenty of details to himself. The British pilots, the flights over Israel, and the various deployments in support of CIA covert operations, for example.

But the PVO had also kept some secrets. A short statement from Major Voronov, the battery commander at Sverdlovsk who had first engaged the U-2, was read to the court. It "confirmed" that a single missile had been fired. Of course, nothing was said about the other missiles, or the downed MiG-19.

After Powers had testified, Soviet technical experts described the U-2's B-camera and SIGINT systems. They gave considerable detail on Systems 3 and 6...the frequencies covered, the antennas, the tape recorders. (Forty years later, the U.S. still considers this information classified!). More experts told of the destructor and the poison pin. Then came the closing speeches.

Powers declared that he was "deeply repentant and profoundly sorry." They sentenced him to ten years confinement, with the first three years in prison. Predictably, his apology wasn't well received by the armchair critics back home. But the State Department announced that Powers had fulfilled the terms of his contract, and would be paid while in prison.

Misunderstanding

The press wouldn't leave the shootdown alone. The New York Times ran a story quoting Powers' father—who had attended the trial—as believing his son had not been shot down. The KGB re-interrogated Powers, and asked him to write a letter of clarification to the paper. "I'm sure my father misunderstood what was said during the trial," Powers wrote. "I hope this letter will clear up any misunderstanding."

Fat chance! In the September issue of True magazine, Drew Pearson and Jack Anderson revealed the "Inside Story of Pilot Powers." This repeated the flameout story in great detail, and added spicy new details. They included how the pilot had made a distress call which was "heard across the Turkish border 1,200 miles away where a handful of Americans were watching the drama helplessly on radar screens and listening by high-powered radio monitors."[53]

But a small group within the CIA knew better. The details about the shootdown which Oleg Penkovsky had passed to the American tourists were at odds with the information which the USSR revealed at the trial. And they were embarrassing to the Soviet regime. This helped the DDP's Soviet Division conclude that Penkovsky's approach was genuine. They laid plans to renew contact with this very promising new source of intelligence.

The Soviet Copy

The remains of the U-2 were removed from Gorkiy Park and analyzed in detail by Soviet aeronautical engineers. The J75 engine excited particular interest. Its gas generator unit was so good that the Soviets thought it might replace the RD-3M turbojet in the Tu-104 airliner. The OKB-16 design bureau in Kazan was instructed to produce a direct copy of the J75, as the RD-16-75.

Two months later, the Soviet government decided that the structural, technical, and maintenance characteristics of the U-2 made it worth copying the airframe itself. The wreckage of Article 360 was carefully sorted and transported to the Beriev Design Bureau in Taganrog. Beriev's team was ordered to produce five copy aircraft designated S-13, and to have the first ready to fly in early 1962. Dozens of other state enterprises were involved in the effort, including one which would reproduce the B-camera, another the ELINT system.

By April 1961, Beriev had completed a full-scale metal mockup of the S-13 fuselage. But it was a huge task reproducing not only the airframe, but all the avionics, fuel system, life support system, and so on. Not surprisingly, weight control was another problem. The effort continued, however, and test rigs for the hydraulics, the autopilot, and many other systems were built. TsAGI (the main Soviet aviation research laboratory) carried out wind tunnel tests, and a Tu-16 bomber was prepared as a testbed for the RD-16-75 engine.

Although the Soviet aircraft industry learned much from the S-13 project, it was never completed. On 12 May 1962 the government in Moscow called a halt.[54] Meanwhile, the Yakovlev Design Bureau had adapted the Yak-25 fighter-bomber for high-altitude reconnaissance, with long, straight wings. It could reach 65,000 feet. Many years later, the Myasischev Design Bureau produced the M-17, a specialized high-altitude interceptor. Both aircraft were sometimes characterized in the West as "U-2skis." But there was one huge difference. The Yak and the M-17 were both twin-engine aircraft. There was nothing to match the real U-2—the only single-engined subsonic airframe which could reach 70,000 feet and stay there for hours on end.

Notes:

[1] Mikhailov and Orlov article

[2] Mikhailov and Orlov article

[3] Mironov article, p6-7

[4] Lockheed Report SP-179, p188

[5] Powers book, p81

[6] Lt Col V. Samsonov, "Eight Missiles for Powers", Motherland (the magazine of the PVO Moscow Air Defence District) 30 August 1992; Col Mikhail Voronov, "The Shooting Down of Powers", abridged from Komsomolets Kubani newspaper in Sputnik: Digest of the Soviet Press, April 1991. Samsonov was in the district command post on 1 May, and Voronov in the first missile battery to engage the U-2. In another article ("U-2: How It Was", HBO - Independent Military Review, No 28, 1998), Igor Tsisar (who was an officer in Voronev's battery at Sverdlovsk) alleges that the radar officer at the Chelyabinsk battery was drunk, and the soldier who was supposed to check the fuses was incompetent.

[7] from an article in Popular Science, August 1994, p58

[8] Mentyukov later speculated that his radar screen had been affected by interference from his own, tail-mounted ECM (electronic countermeasures) system. However, it is possible that the small 'Granger' ECM box on Powers' U-2 really did work as hoped against airborne intercept radar! (In Rich and Janos, p162, Rich notes that Kelly Johnson suspected after the shootdown, that the Granger may have acted in the opposite manner to that intended eg as a homing device for the SAMs that were subsequently fired at Powers. There is no evidence to support this speculation and (sadly) much of the rest of the U-2 story as related by Rich and Janos is inaccurate)

[9] Lt Col Anatoly Dokuchaev, "Duel in the Stratosphere - How the Spy Flight of Francis Powers was stopped," Krasnaya Zvezda (Red Star) 29 April 1990; Dokuchaev, "Flap of the Overflight Mission", Air Fleet Herald, Moscow, November-December 1998. In an interview with Trud newspaper in October 1996, Mentyukov claimed that he managed to reach the U-2's altitude, and as he overtook it, the U-2 was upset by wake turbulence from the interceptor. This led to the spyplane's break-up and descent. This claim is highly suspect, and Mentyukov did not repeat it when interviewed again for the Air Fleet Herald article. The author prefers the more convincing evidence offered by the 'official' Red Star version cited in this footnote, Powers' own account, and other non-official Russian accounts which are cited here.

[10] Voronov and Tsisar articles, and Penkovsky debrief, as above, where the spy notes that "Powers would have escaped if he had flown 1 or 1.5 kilometers to the right of his flight(path)." Tsisar says that he fabricated his official report, as to why two of the battery's three missiles didn't fire. The story that they were prevented from firing because the antenna on the battery's radar cabin obstructed their line of fire, is not correct, he says. By reporting this as fact, he presumably hoped to draw attention away from the battery's slow reaction to the target, which caused the U-2 to be out-of-range by the time that the second and third missiles were ready to launch

[11] Powers book

[12] Zaloga book; Penkovsky debrief, NSA 315, p10; Mikhailov, interviewed by Indigo Films for The History Channel, 1999. Mikhailov estimated that the warhead exploded 20 meters behind the U-2.

[13] Powers book

[14] Tsisar article

[15] Samsonov and Tsisar articles

[16] Boris Ayvazyan, interviewed by Indigo Films for "Mystery of the U-2", a film documentary for The History Channel, 1999. In this detailed account, Ayvazyan contradicts earlier-published accounts (Dokuchaev, Samsonov articles), that he saw Shugayev's missile(s) coming, and instigated his dive without instruction from the ground controllers.

[17] Samsonov article

[18] Pedlow/Welzenbach, p178

[19] Mikhailov and Orlov article, Powers book, Khrushchev book

[20] according to Beschloss p38, the NSA duty officer also sent word of the incident to his opposite numbers at State, FBI, USAF, Army and Navy, although Bissell had stipulated that such intelligence would be sent only to DPD, to preserve U-2 project security.

[21] Beschloss, p32-33

[22] In a telephone conversation with DCI John McCone, transcript published in Beschloss, p404, Eisenhower named Dulles and Bissell as the ones who informed him that the pilot would not survive. Dulles, the 'gentleman spy' was not a technical man, and may thus have believed this himself. Bissell was not technically trained either, but his reputation for assimilating such detail was awesome. He therefore had less excuse for misleading the President. Equally, as a former soldier, one might have expected Eisenhower to have queried the assertion.

[23] Mikhailov and Orlov article

[24] According to MiG-19 flight leader Ayvazyan, interviewed for the History Channel (see above), the truth is more complicated than that. As flight leader, he was supposed to keep his IFF turned on, while his wingman turned his off. But during their refueling at Koltsovo, the two MiG pilots swopped aircraft. After the pair took off to intercept the target, there was confusion as to which aircraft was 'squawking' the IFF code. Ayvazyan turned his on, then off again, since it seemed (to him) that the ground controllers *were* seeing a squawk, from his wingman. Ayvazyan surmised that this caused the ground controllers to mistake him for the target and his wingman as the pair of MiGs in pursuit (as we have seen, Safronov fell some distance behind Ayvazyan, which made this mistake more likely). According to this theory, when the pair were told that the target was "only 5 km away", the 'target' was actually Ayvazyan's (non-squawking) aircraft. As he approached Koltsovo, Ayvazyan says that he did warn ground controllers that his IFF was turned off. But it seems that the missile controllers (in a different room, as noted already) were not informed. They continued to 'tag' Ayvazyan as the intruder and eventually Shugayev's battery was cleared to fire.

[25] Khrushchev book

[26] see Samsonov and Tsisar articles. The author has not been able to determine why the account of 14 missiles being fired, gained widespread currency. It seems to this author and his advisers, including a Russian air defense source, that the Samsonov article ("Eight Missiles for Powers") is the most detailed and reliable of the Russian accounts which have been published.

[27] The standing instructions to pilots before an overflight mission at this time were that they could "tell the full truth about their mission with the exception of certain technical specifications of the aircraft."

[28] Powers, p 305-6

[29] Many contemporary accounts claim that this was a fake, but this is incorrect, according to AUTOMAT photo-interpreter Bill Crimmins. He recognized the airfield in Khrushchev's photo by the row of elderly Tu-4 Bull bombers lined up there. Marty Knutson's flight ten months earlier had flown over the same airfield, south of Sverdlovsk. The Soviets had quickly recovered and developed the B-camera film. However, they had not yet understood the camera's unconventional design, which required the film to be printed emulsion side-up. The photograph which was given to Khrushchev for display, had been printed back-to-front!

[30] Dick Leavitt, Ed Dixon notes and interviews

[31] According to later reports and various Soviet memoirs, Khrushchev did not expect Eisenhower to admit unequivocally that he had authorized the U-2 flights. After he did, though, it is said that the Soviet leader was unable to continue dealing with him, because of the loss of face this would have entailed

[32] reaction of Pakistan and Iran from Beschloss, p268

[33] Tamnes book. According to Beerli, interview, Evang was under such pressure, that DPD's Jim Cunningham even considered 'evacuating' him to the US and arranging a new career for him there.

[34] Robbie Robinson interview

[35] Hansard 11 May 1960, col 398. The next day, suspicious MPs tabled more questions to the Prime Minister. Replying on MacMillan's behalf, Home Secretary 'Rab' Butler told them that "it has never been our practice to disclose either the nature or the scope of intelligence activities" (Hansard, 12 May 1960, cols 619-622)

[36] Brugioni, p47-48

[37] Memorandum of Discussion at the 445th Meeting of the National Security Council, 24 May 1960, as reproduced in Foreign Relations of the US (FRUS) 1958-60, Volume X, document 153

[38] According to Powers book, the delay was set for 70 seconds.

[39] memo of NSC discussion, FRUS volume as above

[40] Memorandum of Conversation, 26 May 1960, FRUS volume as above, document 154

[41] Executive Sessions of the Senate Foreign Relations Committee, Vol 12, SuDocs Ref Y4.F 76/2: Ex3/2/v.2

[42] In fact, these headlines were from one of the more informed reports, in the St Louis Post Dispatch on 7 June 1960, p10

[43] The quotes are from Kelly Johnson's diary, 24 June 1960. The report itself was SP-173, "Engineering Investigation of Airplane 360 Mission Failure", 21 June 1960, declassified and made available to the author, 1999.

[44] White House Memo for the Record, 2 June 1960

[45] "Reconnaissance Flights 1956-57" AIR19/827 file in PRO (the file is mistitled: it actually covers 1960 as well)

[46] Hansard, 12 July 1960, col 1171

[47] AIR19/827 as above

[48] from the USAF Oral History program interview with General Twining, date unknown, p134-5. This may be the basis for the story in Rich and Janos, p158, that Twining had reviewed the last few overflight plans in advance and called Allen Dulles to personally urge changes

[49] author interview with Bill Burke

[50] Powers book, p151

[51] Jerrold Schecter and Peter Deriabin, "The Spy Who Saved the World", Charles Scribner's Sons, New York, 1992 p6-7. The author does not believe that as many as 14 missiles *were* actually fired - see footnote 26

[52] "The Trial of the U-2", Translation World Publishers, Chicago 1960

[53] True, The Man's Magazine, September 1960, p76

[54] S-13 details from Nikolay Yakubovich, "Turnover Spy" in Kryl'ya Rodiny, Moscow, October 1996

Epilogue

On 10 February 1962, Frank Powers walked to freedom across the Glienicker Bridge which separated Potsdam from West Berlin. Walking in the other direction was Colonel Rudolf Abel, the Soviet spy for whom the U-2 pilot was being exchanged. Before the swap was made, Joe Murphy, a CIA security officer from Project HQ, walked to the Potsdam side to make a formal identification of Powers.

Powers underwent a long debrief at Agency "safe" houses near Washington. Then he was taken to see Allen Dulles, who told the pilot "we are proud of what you have done." But Dulles was no longer the DCI. The Kennedy administration had appointed John McCone, a millionaire industrialist, to replace him. And McCone was not at all happy with the way that Powers had behaved. The new DCI set up a Board of Inquiry to formally determine whether Powers had "acted in accordance with the terms of his employment."

The inquiry was chaired by retired Federal Judge E. Barret Prettyman and held in secret. The Board examined 22 witnesses plus Powers himself. They included 12 CIA officers assigned to or associated with Project CHALICE, the Det B commander Colonel William Shelton, and two of Powers' fellow U-2 pilots. Dr Louis Tordella, Deputy Director of the National Security Agency (NSA) and two of his analysts also testified, on the matter of the SIGINT evidence which appeared to conflict with Powers' account of the shootdown. The Board also considered written evidence, including some that came (they were told) from "a reliable Soviet source who was in an excellent position to acquire this information."[1]

That source was Oleg Penkovsky, who had become the West's best-ever spy from inside the Soviet military heirarchy. Following his initial approach to the West just days before Powers' trial in August 1960, CIA and MI6 officers were finally able to meet with the GRU Colonel in London and Leeds eight months later. He handed them technical details of the V-75 (SA-2) missile, and again described his knowledge of the Powers shootdown. "No direct hits were made...but (the U-2) was within the radius of explosion of one of the missiles...the U-2's tail and wing assembly were slightly damaged," he told them. "The damaged parts were not shown at the Moscow exhibition, and your intelligence personnel should have spotted that," he added.[2]

Powers Cleared

The three-member Board of Inquiry worked intensively and completed the investigation within a week. Their report to DCI McCone dated 27 February 1962 unequivocally cleared the U-2 pilot. They found that he had told the truth about the MayDay flight and shootdown, and conducted himself properly throughout his confinement. The only evidence which conflicted with his version of events came from the NSA intercepts, the Board noted. But Tordella and his analysts admitted "the obvious possibility of confusion and error" within the Soviet air defense system. The Board concluded that it could not "make a flat assumption of accuracy" from the SIGINT reports.[3]

McCone wasn't happy, and neither was the NSA. McCone asked the Board to reconvene on 1 March and reconsider the evidence.[4] Prettyman told the DCI that the NSA's "evidence" was at best unverifiable hearsay once removed.[5] McCone reluctantly accepted the Prettyman report. He took it to the White House, where President Kennedy endorsed it. On 6 March 1962, a sanitized version was released to the public, just before Frank Powers testified to an open session of the Senate Armed Services Committee. The public version finessed the references to the SIGINT evidence, which was still highly classified: "Some information from confidential sources was available..that which was inconsistent (with Powers' own account) was in part contradictory with itself and subject to various interpretations. Some of this information was the basis for...subsequent stories in the press that Powers' plane had descended gradually from its extreme altitude and had been shot down by a Russian fighter at medium altitude. On careful analysis, it appears that the information on which these stories were based was erroneous or was susceptible of varying interpretations." (In a curious replay of history, the danger of relying too heavily on SIGINT intercepts of the Soviet air defense systems was illustrated anew in 1983, when the Korean Air Lines Boeing 747 was shot down over Kamchatka. President Reagan accused the USSR of deliberately destroying a civilian airliner, based on NSA SIGINT analysis of dubious validity).

Senate Hearing

The Senate hearing was Powers' first chance to explain himself properly in public. He did not fluff it. The exact circumstances of the shootdown, the destructor, the poison pin, the cover story, and all the other controversies became a matter of public record. Powers stayed on the CIA payroll for a while as a training officer, helping to teach recruits about Soviet interrogation methods and the like. In September

1962, he was finally allowed out to North Base. Over a three-day period, he flew intensively with Jim Cherbonneaux in one of the CIA unit's T-33s. Cherbonneaux confirmed that Powers had lost none of his flying skills. The visit ended with an alcoholic reunion party with the half-dozen other pilots from Projects AQUATONE and CHALICE who, like Cherbonneaux, were still flying U-2s for the CIA.

Then Powers took up an offer from Kelly Johnson to become an engineering test pilot on the U-2 program. He reported for work at Burbank on 15 October 1962. That very day, the U-2 was proving its continuing worth to U.S. intelligence. Photo-interpreters at NPIC and SAC were discovering Soviet offensive missiles in Cuba, from film exposed on a U-2 mission the previous day.

In the Cuba Missile Crisis, President Kennedy was able to face down Premier Khrushchev because he had solid intelligence on the USSR's strategic military capabilities. His predecessor in the White House never enjoyed such a luxury. And after the U-2 was shot down, and Eisenhower banned further overflights, the U.S. was virtually blind for over a year about Soviet missile developments.

CORONA success

On 19 August 1960 the first recovery of a film capsule from a U.S. reconnaissance satellite was made. It was the ninth attempt in the CIA's top-secret CORONA project. Adding the additional launches for R&D purposes, it was the 14th time that the Thor-Agena combination had blasted off from Vandenberg AFB in California. But the program managers and their political masters, notably President Eisenhower, had persevered despite all the failures.

The satellite had made eight north-south passes over the USSR. But when the film reached Art Lundahl's men at the PIC in Washington, it got a mixed reception. There were the inevitable clouds—only 25% of the coverage was completely cloud-free. Some of the PIs were disappointed by the resolution of the smaller-scale film exposed by the satellite's panoramic camera. At 20-30 feet it was nowhere near as good as that obtained by the large-format framing cameras carried by the U-2.[6] But, said others, look at the coverage! The first successful CORONA mission covered more Soviet territory than all the previous U-2 overflights combined.

The resolution was good enough for the PIs to identify several new airfields and 20 new SA-2 missile sites. Mys Schmidtya airfield on the Chukotsky Peninsula (a target for SAC's peripheral U-2 missions in the spring of 1959 and 1960) was imaged on the second pass. Kapustin Yar and the western end of the Saryshagan impact area also showed up cloud-free. But there was still no imagery of offensive missile sites in the northern USSR. The Polyarny Ural area which was rated the likeliest area for deployment was completely covered by cloud.[7]

More CORONA missions were scheduled, but the next three were failures. On 10 December 1960, the second recovery of a capsule with successfully-exposed film was made. This time, the satellite's orbit had taken it directly over Plesetsk, but high cirrus cloud covered the target area, making positive interpretation impossible.[8] The next successful CORONA mission wasn't until June 1961, and it took two more successes in July and early September before the CIA's missile analysts felt confident enough to make a "sharp downward revision" in Soviet ICBM strength. Their assessment was formalized in a mid-September revision to the 1961 National Intelligence Estimate on Soviet long-range missiles, which had only been issued three months earlier.[9]

The latest CORONA imagery showed three ICBM sites under construction. But they were further south than had been suspected, all below 60 degrees North. Moreover, although the launch complexes were close to rail lines, the launch pads themselves were served only by road. The PIs reviewed coverage of Tyuratam from the U-2 mission 4155 on 9 April 1960, which first showed the new type of road-served launch pad. Then the all-source analysts went to work adding other intelligence, such as SIGINT from the Soviet missile tests from Tyuratam during 1961, and the considerable information which Penkovsky was providing "from the inside."[10] Now, the picture of Soviet missile deployment became much clearer.

Soviet Missiles

The Soviets had curtailed deployment of their first big ICBM (the R-7, or SS-6) in favor of one, possibly two smaller, more flexible missiles. They corresponded to the U.S. Titan ICBM in that they were liquid-fueled but tandem-staged, with the upper stage igniting at altitude. The first generation ICBMs, both Soviet and the U.S. Atlas, ignited the upper stage at launch. There had been a large number of test-launches of the Soviet second-generation missiles in 1961, but including many failures. They would probably not be operational until late 1962.

Yur'ya was the launch site for these new Soviet missiles, where construction was the most advanced. This site was reached by a spur off the railroad leading northwest from Kirov. The analysts concluded that the complex must have been started in the autumn of 1959. Had Frank Powers not been shot down over Sverdlovsk, he would have flown right over Yur'ya about two hours later, as he followed the railroad from Kirov to Kotlas.

And what of Plesetsk, the primary target for the U-2 mission of 1 May 1960? Good imagery of the remote, cloud-covered site in the northern tundra was still eluding the analysts. One of the mid-1961 CORONA missions had provided limited coverage, but showed only the SA-2 sites which protected the complex, several large support areas, and numerous buildings. But the missile analysts had nearly figured it out. They reckoned that the Plesetsk site housed two, four, or more launchers for the first-generation ICBM. Making allowance for some yet-to-be discovered sites, the analysts now reckoned there could be 10-25 of these first-generation

Soviet ICBM launchers ranged against the U.S. As usual, the USAF disagreed. In a formal dissent to the NIE, the blue-suiters alleged that there were about 50 launchers, posing "a serious threat to U.S.-based nuclear striking forces."[11]

In fact, the true number of launchers was...just four, and they were *all* at Plesetsk. Despite all the bombast and the boasts from Premier Khrushchev, the Soviet ICBM program had suffered serious setbacks. The shortcomings of Sergei Korolyev's "Big Seven" as an operational ICBM had led to the canceling of plans to build other R-7 launch sites after Plesetsk. Mikhail Yangel's alternative ICBM, the R-16, was hurriedly advanced. This was indeed the second-generation Soviet ICBM identified by U.S. missile analysts. But on 24 October 1960, there had been a terrible accident on the new launch pad at Tyuratam, while the R-16 was being prepared for its first test flight. A test signal went wrong, and ignited the second-stage rocket engine. The highly unstable mix of nitric acid oxidizer and hypergolic fuel in the first stage was immediately below. A cataclysmic blaze erupted, killing over 200 technicians. Most were totally consumed by the flames. They included Marshall Nedelin, the head of Khrushchev's much-vaunted new Strategic Missile Troops, and most of Yangel's design team.[12]

It was a huge setback for the Soviet missile program. The R-16 did enter flight testing in 1961, as was duly observed by U.S. intelligence. But progress was slow compared with the U.S., which began deploying 60 Titan ICBMs to join a similar number of Atlas ICBMs already on nuclear alert. From 1960 to 1962 there was indeed a substantial missile gap—but it was in favor of the United States, not the Soviet Union!

Notes:

[1] Board of Inquiry report to the DCI, 27 February 1962, declassified with redactions, 1978

[2] CIA transcript of Penkovsky debriefs, 20 and 23 April 1961, declassified IRONBARK materials, made available by the National Security Archive, Washington DC, reference NSA286. The transcript shows that Penkovsky's knowledge of the Powers shootdown was not accurate in every detail, such as the number of missiles fired; his main informant was a Major General who headed the Political Directorate of the PVO.

[3] Board of Inquiry report, as above.

[4] Pedlow/Welzenbach, p185

[5] Thomas Power, "The Man Who Kept The Secrets", Alfred Knopf, 1979, note 4 to Chapter 6, p379

[6] Dwayne Day, in "Eye in the Sky", p61

[7] CIA CORONA documents, p52, 81 and 119

[8] remarks by David Doyle at the CIA's CORONA conference

[9] NIE 11-8/1-61, 21 September 1961, as reproduced in CIA CORONA documents, p130

[10] Penkovsky's information was described as "reliable clandestine reports" in the NIE

[11] NIE as above, p130, 132

[12] Zaloga, Target America, p191, 194-197

Postscript

The "U-2 incident" was a seminal event in the Cold War. Not only did it cause high political drama, it had all the ingredients of a spy novel. Many of those who were part of the U-2 program in those early years felt they had been touched, in some small way, by history. The rest of us participated vicariously, thanks to the many articles, books, television programs, and films which followed.

The first book was published just ten weeks after Powers was released.[1] The authors got no official co-operation, and the CIA discouraged Powers from publishing his own account. John McCone remained hostile to the returned U-2 pilot. In April 1963, the CIA's Intelligence Medal was awarded to the project U-2 pilots in a secret ceremony. Powers was left out. (He eventually received the 10-oz gold star two years later, when McCone left office).[2]

Far from being grounded by the MayDay incident, the U-2 remained in high demand, although it was never deliberately flown over the USSR again. The CIA ran a scaled-back operation throughout the 1960s codenamed Project IDEALIST. The Agency's U-2s ranged over Cuba, Vietnam, and China—the latter in a joint venture with Taiwan and using Republic of China Air Force (ROCAF) pilots. Five U-2s were shot down by Beijing's SA-2 missiles between 1962 and 1967, before overflights of the Chinese mainland were halted. SAC lost one U-2 and its pilot to another SA-2 during the Cuba Missile Crisis, but also found the aircraft too valuable to discard. The Skunk Works was kept busy with routine U-2 maintenance and a never-ending series of updates to the aircraft. These included inflight refueling, electronic warfare equipment, and even a modification to allow landings and take-offs from aircraft carriers. In 1967, an improved and enlarged version designated U-2R flew for the first time, and the government ordered 12, six each for the CIA and USAF.

As a result, Frank Powers found plenty to do in his new career as a U-2 test pilot with Lockheed. He clocked up another 1,400 hours on the aircraft from 1962 to 1969. But after the last U-2R was delivered, the work dried up. In October 1969, Kelly Johnson served notice that his employment would soon be terminated. Meanwhile, Powers decided to write his book after all. The CIA raised no official objection, and Powers requested no official help. The book was published on 1 May 1970, exactly ten years after the shootdown.[3]

Powers' Book

In the closing pages of "Operation Overflight," Powers reviewed some evidence that his MayDay flight might have been "betrayed." This was a recurring theme in the growing body of U-2 literature. It was strictly speculation, the U-2 pilot explained.

He referred to the case of two SIGINT analysts from the NSA who defected to the USSR in July 1960. William Martin and Bernon Mitchell both monitored Soviet air defenses during U.S. airborne reconnaissance operations in the late 1950s, notably the SIGINT flights conducted specifically for the NSA along Soviet borders. In the course of their work, Martin and Mitchell may have gained some insight into the U-2 overflights. They undoubtedly did grave damage to the U.S. by passing on many of the NSA's secrets, beginning in 1958. Powers wondered whether they had learned of his flight in advance, during the course of their duties, and had passed the information to their Soviet handlers.

Then Powers raised the problems in communicating the "go-code" to Peshawar on the morning of 1 May 1960. After his return to the U.S. in 1962, said Powers, he had conversed with Det B's communications supervisor at Incirlik on 1 May. The "commo" man told Powers that there had been a problem relaying the go-code from Germany to Turkey, and that it had eventually been relayed "over an open telephone line." The supervisor told Powers that he was off-duty at the time, and learned of the incident later. Had he been on duty himself, the supervisor said he would never have sent the message on to Pakistan, since "the risk of the call having been monitored (was) so great."

Finally, there was Lee Harvey Oswald. The man who assassinated President Kennedy served in the U.S. Marine Corps at NAS Atsugi, Japan, for 14 months in 1957-58 as a radar operator. A year later, he "defected" to the USSR, and remained there until 1962. During his tour at Atsugi, Oswald undoubtedly witnessed Det C's U-2 operations at close quarters, and may have learned something of its high-altitude capabilities from the height-finding radar that he helped to operate.[4] Powers suggested that Oswald may have revealed the U-2's altitude to the USSR, six months before he was shot down.

Speculations

As speculations go, these by Powers in his autobiography were reasonable. Others were not. For example, in a 1975 television documentary, Norwegian fisherman and convicted Soviet spy Selmer Nilsen claimed that the U-2 had been downed by a bomb planted in its tail before takeoff. In a 1993 book, former KGB Colonel Victor Sheymov recounted a conversation that took place during the 1970s with a Major in the Russian air force. According to Sheymov, the officer claimed that the USSR "knew the exact flight plan of every U-2 invading Soviet airspace several days in advance." This information was provided by a Soviet source "who was also able to influence" the flight plans. The source had managed to fix things so that Powers "flew within reach of the ground-to-air missiles" on 1 May 1960.[5]

What are we to make of these stories? Nilsen's bomb can be easily dismissed. At least half of the 20-strong launch crew in Peshawar would have to have been party to the planting of any bomb. Sheymov's spy is difficult to disprove: such is the nature of cloak-and-dagger allegations. However, Sheymov was himself a spy, exfiltrated from the USSR in 1980 by the CIA. There is no indication that the CIA took the allegation seriously, if indeed Sheymov even suggested it to the Agency during his debriefings. And the ancillary detail provided by Sheymov, about the Soviet reaction to U-2 overflights, is far from convincing.

Might the NSA analysts Martin and Mitchell have tipped off the Soviets about impending U-2 flights, as Powers wrote? According to senior U-2 project officers consulted by this author, the NSA was *not* informed in advance of any U-2 over-flight plan.[6] In any case, the two NSA analysts were in no position to influence flight routes.

The "Go-Code"
Much has been made of the unusual circumstances in which the "go-code" was relayed to Peshawar. Could the USSR have intercepted the signal, and thus been alerted that the flight was about to take place? The account that Powers gave in his book differs in material fact from that recorded in the CIA's archives, and related in a previous chapter of this book.[7] However, it is true that the correct procedure for transmission was *not* followed at some stage of the message's journey onwards to Pakistan from Germany.

It should be noted that the go-code alone would have told the Soviets nothing about the route of the impending flight. It was only the "execute" signal. The mission plan itself (giving the route co-ordinates and intelligence objectives) was transmitted much earlier (routinely, it was 12 hours before a flight, but in this case, some days earlier since the mission was postponed twice). The mission plan would have been only one of several signals relayed each day from Project HQ to the U-2 detachment. Some of these dealt with routine matters, while others had no purpose at all. This was to ensure that no pattern of communication could be discerned by eavesdroppers, enabling them to make deductions about pending U-2 operations.[8]

Despite this precaution, we now know that some messages to Det B were monitored by the Soviet Union, although they may not have been decoded. Prior to the overflight on 9 April 1960, Soviet communications intercept facilities in the Transcaucasus region did pick up indications that a special operation might be imminent. However, the subsequent Soviet enquiry into the failure to shoot down that flight found that this piece of intelligence was not reported to the command element because of a number of chance happenings.[9]

Moreover, the Soviets might have been alerted to the flight through their own efforts on the ground. Despite the precautions that were taken, it would have been possible for KGB agents to observe the U-2 operation at Peshawar. The taxiway to the runway ran close to the perimeter fence of the airfield. At least one pilot of an earlier overflight mission from the Pakistani base recalled seeing bystanders outside that fence as he taxied out.[10]

Furthermore, the hangar which was used by Det B at Peshawar was not secure during normal working hours, according to one member of the Det B deployment crew. Various Pakistani base workers visited, and there was even a tuition class for schoolchildren held there one day, during the crew's long wait to launch the mission.[11]

Soviets Alerted?

The lack of security at the hangar may have been the reason why it was decided to ferry the U-2 to and from Peshawar, rather than keep it hangared there during the daytime. But that decision itself may have led to a security breach, according to Colonel Stan Beerli, the former Det B commander who was waiting in Bodo for Powers to arrive. Beerli swears that he knew nothing of the decision to ferry the U-2 to and fro, if the mission should be postponed. He would not have approved such a policy for fear of alerting the Soviets. Beerli notes that their early warning radars could have picked up the aircraft as it was being ferried across Iran and Afghanistan, and the more such flights there were, the more likely a radar intercept was.[12] Certainly, we know now that the latest PVO early warning radars in this area could "see" more than 200 miles past the Soviet border; even Peshawar itself is inside the 200 mile radius, although the Hindu Kush and Pamir mountain ranges would have obscured the radar line-of-sight at lower altitudes.

The "shuttle" flights, then, might have helped the Soviets deduce that an overflight was imminent. Conversely, though, U-2s had been performing regular, non-penetrating SIGINT flights along the Soviet border. Indeed, for the last few missions over the USSR, it became the practice to deliberately launch a "diversionary" U-2 flight along the southern border from Incirlik, at about the same time that the overflight aircraft took off from Peshawar.[13]

The Russian sources consulted by this author do not suggest that the PVO was pre-alerted that a U-2 would be penetrating their airspace on or just before 1 May 1960. In the final analysis, indeed, the Soviet air defense system should not have needed such a warning. The PVO had been thoroughly "tipped off" three weeks earlier, by the previous U-2 overflight. That mission was easily detected and tracked as soon as it crossed the border. When the PVO failed to intercept the intruder, all hell was raised by Premier Khrushchev. The PVO's state of readiness was further improved after the 9 April incident. Ironically, the MayDay holiday caused it to be somewhat relaxed again, and this evidence counts against the possibility of a specific tip-off.[14] (Incidentally, it has been suggested that the flight was launched on this very day, precisely *because* U-2 mission planners hoped to take advantage of just such a relaxation. As we have seen, however, weather conditions and President Eisenhower's deadline were the two crucial factors governing the launch date).

What is certain, is that the Project HQ was well aware that the 9 April flight had been detected and tracked from an early stage by the PVO. That did not stop Bissell and Dulles from recommending another flight. With the considerable benefit of hindsight, we can judge that they were over-confident, and made the wrong call. But so did Premier Khrushchev. If he had swallowed his pride and protested the 9 April flight, all the evidence suggests that President Eisenhower would have canceled Operation GRAND SLAM. And there would have been no "U-2 incident" at all.

Death of Powers

After leaving Lockheed and having his book published, Frank Powers was unemployed for a while. He was understandably upset when his application to join three former colleagues and U-2 pilots flying U-2s under contract to NASA was turned down for political reasons. Then he started flying for a Los Angeles radio station, reporting on traffic conditions from the air in a Cessna lightplane. The book was turned into a film, starring Lee Majors. In late 1976 Powers was hired by television station KNBC and sent on a helicopter flying course. He would be piloting the station's heavily-equipped "telecopter" to provide live pictures of fires, police chases, and other newsworthy events in the L.A. area.

On 1 August 1977 Powers and his cameraman were returning to Burbank after covering a brushfire near Santa Barbara when the engine quit. The helicopter crashed onto a sports field in Encino, killing Powers and his passenger. It carried a TV camera and data-link antennas on unweildy external mounts, which may have contributed to Powers' failure to control the machine's unpowered descent. Some suggested that Powers did not have enough helicopter flying experience to cope with such an emergency.

Whatever the case, it seemed an inappropriate end for the man who had cheated death over the Soviet Union. Moreover, there was a sad irony to come, when the official accident report on the helicopter crash was released. Having flown 2,000 hours in the U-2, where fuel management was one of a pilot's highest priorities, Powers had allowed the helicopter's fuel tank to run dry![15]

Notes:

[1] David Wise and Thomas Ross, "The U-2 Affair", Random House, May 1962

[2] Pedlow/Welzenbach, p186. Ironically, none of the pilots was allowed to retain the Medal when it was first awarded. Two years later, this decision was rescinded for those pilots who were no longer flying for CIA. In fact, the first Medal given for retention in 1965 was the one given to Powers. Powers never did receive the USAF's Distinguished Flying Cross, which was awarded to all his contemporaries in the U-2 program in 1968. But in 1987, Leo Geary finally persuaded HQ USAF to correct the wrongdoing. The retired Brigadier-General spent ten years from 1956 to 1965 as the Pentagon's chief liaison officer for the CIA U-2 program. He presented the medal to Powers' second wife and widow, Sue, at a reunion of U-2 and A-12 veterans in Las Vegas.

[3] Beschloss, p398, suggested that Powers lost his job at the Skunk Works because he decided to write the book (and that the CIA had been paying his salary all along, via Lockheed). In truth, Powers was no longer of any use to Lockheed. He had never gone through test pilot school, and was therefore not qualified to test-fly any other Lockheed airplanes. Further, Kelly Johnson's log for 25 September 1969 noted that "he doesn't have an ATR (Air Transport Rating), so we have no other job for him - not even flying the Beechcraft."

[4] In "Oswald and the CIA", Carroll and Graf, 1995, author and conspiracy theorist John Newman devotes an entire chapter to a confused and inaccurate account of Det C operations, what Oswald may have known, what the CIA knew about what Oswald may have known, and so on.

[5] Victor Sheymov, "Tower of Secrets", Naval Institute Press, 1993, p137

[6] Burke, Beerli interviews

[7] Powers refers to the go-code being transmitted from Germany to Turkey by open telephone line, following problems in passing the signal by the usual radio means. CIA records show that the difficulty in transmission was between Turkey and Pakistan, and the solution was to use a different HF frequency (Pedlow/Welzenbach, p175-6)

[8] author's interviews with CIA 'commo' operators.

[9] Col Alex Orlov, remarks posted to the website of the CIA/Center for the Study of Intelligence, after the 1997 conference. In "The Samson Option", as above, p51 note, author Seymour Hersh asserts that the USSR *did* deduce when U-2 overflights were scheduled, from message traffic analysis. Hersh says that USAF communications technicians themselves monitored "the extensive and poorly masked preflight communications between Washington and the U-2 airfields." Hersh gives no sources for this information. Hersh further alleges that the USSR routinely grounded all their air traffic whenever a U-2 was overflying , to make the task of detection and tracking easier. This did indeed happen on 1 May 1960, but this author has not been able to verify that it happened on any other occasion

[10] author interview

[11] author interview with Jim Woods. On 17 May 1960, the US embassy in Kabul was informed by the Afghan foreign minister that during his wait at Peshawar, Powers had been "entertained socially by his Pakistani officer opposite numbers, who knew all about his mission." Since the foreign minister's source for this information was the USSR, this author believes this was disinformation, and gives it no credence

[12] author interview with Stan Beerli

[13] Pedlow/Welzenbach, p163

[14] According to Mikhailov/Orlov (article), the PVO stood down a V-75 (SA-2) regiment at the prime target of Tyaratam for the holiday. Why would they have done that, if they knew for certain that a U-2 overflight was about to be launched?

[15] During the accident investigation, it was revealed that the helicopter had developed a faulty fuel guage some days earlier. Powers had written up the fault. There was some evidence to suggest that a mechanic had fixed it without signing the repair off. His failure to do so might have led Powers to believe that he had a few more gallons remaining, than was actually the case.

Sources and Acknowledgements

Following publication of my first book on the U-2 in 1989,[1] various people formerly or currently associated with the U-2 program were generous enough to praise my efforts. It was their encouragement which eventually motivated me to return to the subject in a book-length treatment. In the intervening years, I have been privileged to meet many more U-2 veterans, and also to fly in the aircraft itself, courtesy of the U.S. Air Force. By default rather than design, I have become the "corporate memory" on U-2 history and an unofficial source on the aircraft's current operations. It's been great fun—and a great honor.

My main sources for this book have been interviews and declassified documents, together with those books and articles published mostly in the last decade, which modified or supplemented my previous knowledge of the subject. These sources are referenced extensively in the footnotes to each chapter. I particularly thank the interviewees, and regret that they cannot all be listed here.

The declassification process has been long and unsatisfactory. Despite the passage of time, virtually no primary documents that were generated within the U-2 program and originally classified above the Secret level, have been released. This has complicated the task of analyzing political or operational direction, but the release of some relevant White House and State Department materials mitigated the problem somewhat. The CIA promised much but released little, except a U-2 program history written in the 1980s (with redactions).[2] Some key British files are still retained. The Soviet side of the story is told mostly through memoirs published by some of the key participants.

A special vote of thanks is due to Joe Donoghue, whose interest in the U-2 was sparked by a two-year assignment to the program in the 1960s. His own diligent efforts to preserve (and recreate!) the history of the program have been invaluable, and he provided direct research assistance and much-needed encouragement to this author on numerous occasions.

The following rendered special assistance: Col Stan Beerli (USAF, retd) whose credentials are explained in the Foreword to this book; Dino Brugioni and Bill Crimmins, both formerly of PID and PIC; Bill Burr at the National Security Archive and Dwayne Day at the Space Policy Institute, both housed within George Washington University; Kay Cherbonneaux and Bob Tripp; Nigel Eastaway of the Russian Aviation Research Trust; Linda Garmon of WGBH-TV in Boston; Yefim Gordon, who provided and translated Russian documents; Cargill Hall, formerly with the USAF History Office and now the chief historian at the NRO; Brig Gen Leo Geary (USAF retd), former Pentagon liaison officer to the CIA U-2 project; Maj Gen Pat Halloran (USAF retd) former U-2 and SR-71 pilot and wing commander; Ernie Joiner, Bob Klinger, Bob Murphy, Jim Wood and Dave Young, former employees of the Lockheed Skunk Works; Bob Dunn, Denny Lombard, and Garfield Thomas, current employees of the Lockheed Martin Skunk Works; Jay Miller, whose own books on the U-2 remain significant sources of reference; Gary Powers Jr; Jeff Richelson; Mick Roth; Don Welzenbach, former CIA historian; Col Chuck Wilson, former U-2 pilot and squadron commander; Dave Wilton of the British Aviation Research Group; and Dr Jim Young at the AFFTC History Office.

My thanks are also due to the Norwegian Aviation Center for inviting me to speak at their Cold War Forum at Bodo in October 1995, and to the CIA's Center for the Study of Intelligence for a similar speaking invitation to their conference "The U-2: A Revolution in Intelligence" in Washington, D.C., in September 1998.

I have not received any advances, research grants, or other financial assistance while writing this book. I am therefore most grateful to those kind folks who provided accommodation or other help which reduced my expenses. They included Reuben Johnson, Dave Klaus, and Ash and Sally Lafferty in Washington; Tony Bevacqua and Bill Bonnichsen in California; and Linda Peri in Colorado.

Last but not least, I must acknowledge the patience and understanding of my wife, Meng. Without her support, you would not be reading this volume today.

Notes:
[1] "Dragon Lady - The History of the U-2 Spyplane" by Chris Pocock, Airlife Publishing, UK and Motorbooks, US, 1989. This book is no longer in print.
[2] "The CIA and the U-2 Program, 1954-1974", by Greg Pedlow and Don Welzenbach, CIA History Staff, declassified 1998. These authors had access to operational U-2 files generated by the CIA, but were not able to conduct an extensive interview program to supplement or modify this data. Also in 1998, an inter-agency declassification panel finally authorized the release to the National Archives, of imagery taken on many of the early U-2 flights. At the time of writing, however, this imagery remains virtually useless to historical or environmental researchers, since no meaningful positional index has been released.

Appendix 1:
Technical Data

Airplane Maximum Performance Summary

Model	MTOGW lbs	Fuel gals	Range naut miles	Flight Time hr:min	Altitude feet
U-2A (J57-P-37 engine)	20,080	1,335	3,775	9:00	72,000
U-2A (J57-P-31 engine)	19,665	1,335	4,175	10:00	72,500
U-2C (with slipper tanks)	23,040	1,545	4,600	11:30	75,900
U-2C (w/o slipper tanks)	21,600	1,345	4,150	10:45	76,400

Notes: MTOGW = Max Take-Off Gross Weight. Range values are to zero fuel. Data is derived from Lockheed Flight Test Reports SP-109 (U-2A) and SP-179 (U-2C), declassified 1999, supplemented by author's interviews. Since different flight profiles were required to achieve maximum altitude and maximum range/flight time, the values shown above for these categories could not both be achieved on the same mission. Also, the maximum values shown are those that were used for flight planning purposes. Some higher value were achieved on test flights.

Airplane Basic Dimensions

Wing area	600 sq ft
Wing span	80 ft
Aspect ratio	10.67
Fuselage Length	49 ft 8 in
Height at Tail	15 ft 6 in

Engine ratings

Pratt & Whitney model	Dry Weight lbs	Thrust lbs	Max EGT deg.C	Max P.R.
J57-P-37	4,050	10,500	640	3.07
J57-P-31	3,615	11,200	610	3.24
J75-P-13	4,900	17,300	630	3.3

Notes: Data is from Lockheed Flight Test Reports SP-109 and SP-179. Early examples of the J57-P-31 (YJP-31) weighed 3,675 lbs.

Imaging Sensor Configurations

A-1

One Hycon HR731 camera with 24-inch focal length on rocking mount to provide left, right and vertical options, plus HR732 film magazine, format 9 x 18 inch x 1,800 feet

Three Hycon HR730 cameras with 6-inch focal length in fixed trimetrogen mount, film format 9 x 18 inch x 390 feet

A-2

Three Hycon HR731 cameras fixed at vertical, 37 degrees left and right, plus HR732 film magazines

B

One Hycon HR73B1 camera, 36-inch focal length, f10, shutter aperture 3.6 inches, with seven 'stop-and-shoot' positions (left and right 1,2,3 plus vertical). Four modes of operation (1 - all positions; 2 - L1, R1, vertical only; 3 - L1,2,3 plus vertical only; 4 - R1,2,3 plus vertical only). Normal stereo overlap 55 degrees. Film magazine contained two contra-winding 9.5 x 18-inch rolls x 4,000 or 6,500 feet, to obtain film format of 18 x 18-inch. Flight weight 484 lbs (2 x 4,000 feet film load) or 577 lbs (2 x 6,500 feet film load)

C

One Perkin Elmer 180-inch folding focal length camera. One example developed and issued to 4080th SRW, and three aircraft modified to receive (Article 343 for CIA, Articles 379/56-6712, 388/56-6721 for 4080th SRW) but not used operationally

Tracker

One Perkin Elmer 3-inch focal length panoramic camera, f8, film format 70mm x 9.5-inch x 1,000 feet. Weighed 55lbs when film loaded.

D

Westinghouse APQ-56 side-looking radar, Ka-band, plus ASN-6/RADAN system for improved navigation accuracy

SIGINT Sensor Systems

NOTE: since no data on the U-2's SIGINT systems has been declassified, this information is provisional, and based on various previously-published sources, U-2 service bulletins, and author's interviews

System 1

Ramo-Woolridge (later STL) ELINT receiver, covered E/F (old S), G/H (old C) and I/J (old X) bands. D- (old L-) band possibly added later. Crystal video receivers. Provided PRF and scan rate (not frequency). Needed to be preset to one or other of these bands before takeoff? Recorder was under right console in cockpit.

antennas and receivers in nose, used the square shaped radomes. Pilot selected left, right or both antennas. Was a standard equipment from the earliest deployments on both CIA (till replaced by System 6 in 1958?) and SAC aircraft (except hard-nose samplers).

System 2
This number appears to have been allocated to the HF navigation system, development of which was initiated in 1955, but soon abandoned.

System 3
Ramo Woolridge/STL COMINT receiver, covered VHF (100 - 156 MHz) band. Search, lock and record. The receiver and recorder and antenna were all originally in nose, latter using long radome beneath forward nose. From 1958 shared flat scimitar-type antenna inside lower ventral radome with System 6 on CIA aircraft. The three swept-frequency receivers plus recorder (with one channel for each receivers) were moved to left cheek on CIA aircraft at same time. Was a standard fit on CIA and SAC aircraft and (together with System 1) could be carried on camera-equipped aircraft.

System 4
Ramo-Woolridge/STL multiband SIGINT receiver, covered 150MHz-40GHz, occupied entire Q-bay, weighed 570lbs. Available by early 1957. Possibly used only by SAC on its four dedicated ferret aircraft. See System 5.

System 5
Possibly the equivalent of System 4, used by CIA only.

System 6
STL? miniaturized wide-frequency receiver. Replaced System 1 on CIA aircraft only from 1958, and may have originally used some of its antennas/radomes. A pair of G/H (old C)-band antennas were housed behind square fibreglass panels on either side of the nose. A pair of D (old L)-band antennas (helical spiral of flat printed circuit) behind round fibreglass panels at front of Q-bay, with pair of IJ (old X)-band parabolic antennas immediately to rear behind small square panels. In addition, VHF (old P)- band (50 - 150 MHz) antenna was encased in the lower ventral fairing, being shared with System 3. Control panel in cockpit to select L or R antenna (as appropriate, for border surveillance) or BOTH (for overflights). Panel also displayed two lights which indicated that antenna switching was taking place.

When BOTH selected, the coax antenna switches cycled every 30 seconds to give alternating coverage to both sides of the flight path.

Two three-channel recorders working at 2 1/4 inches per second: one for C and X-band plus time code/antenna switching was located under rear of right cockpit console; another for L and P-band was in the fuselage aft of the engine intake, behind the right cheek access hatch. These recorders were three-track, with time code and left-or-right indication on the centre track

The X-band pre-amp was mounted on the front of the Q-bay; small crystal detectors and amplifiers for the other bands were in the right cheek and nose.

System 7
Possibly a dedicated missile telemetry intercept system used by CIA at Det B. It used the upper rear fuselage ('ram's horns') antennas which were originally intended for System 6, but replaced on that system by the single lower rear fuselage ventral antenna. Most equipment was in the Q-bay. Pilot's comments could be added to the (very high speed?) recorder.

Sampling Systems
A-Foil
first sampling system consisting of four 16-inch diameter filter papers rotated through duct leading from small opening at the tip of the aircraft's nose, to capture air particles. Six U-2A-1 models for 4080th wing only.

F-2 Foil
second sampling system, also to capture air particles. Used six of the 16-inch filter papers fed by air from intake duct faired onto left side of Q-bay hatch. Issued to CIA and 4080th SRW, used in conjunction with:

P-2 Platform
also known as the 'ball sampler', consisting of six 13-inch diameter spherical receptacles for air (gas) sampling. Each had 944 cu inch capacity, and was filled with air compressed to 2,000 psi from the engine bleed.

Other Payloads
Ivory Tower I/II
dedicated hatch-mounted weather reconnaissance payloads.

US Mule
dedicated hatch for leaflet dropping. CIA only. never used operationally?

Appendix 2:
CIA Organization and Key Personnel

Note: most of the information below has been compiled by the author from interviews; no detailed organizational records have been declassified. Some titles, dates and functional relationships have been assumed, and some relevant personnel may not be listed.

Project HQ
"Development Project Staff" (DPS) established December 1954 reporting to the Director of Central Intelligence (DCI). Became part of the new Development Project Division (DPD) in February 1959, reporting to the Deputy Director for Plans (DDP).

Richard Bissell	Special Assistant to DCI for Planning and Co-Ordination (SA/PC/DCI), December 1954>
	Deputy Director for Plans (DDP) January 1959>
Col Ozzie Ritland	military deputy to SA/PC/DCI, August 1955>
Col Jack Gibbs	military deputy to SA/PC/DCI, May 1956>
Col Bill Burke	military deputy to SA/PC/DCI, May 1958>
	Acting Chief, DPD, January 1959>
Herb Miller	administrative director, 1955>
Jim Cunningham	administrative director, 1956>
Col Marion Mixson	director of operations, 1956>
Col Paul Gremmler	director of operations, 1958>
Col Stan Beerli	director of operations, August 1959>

U-2 Test Unit
established at Watertown Strip (Groom Lake) 1955, moved to North Base, Edwards AFB, June 1957
Col Dick Newton	base/unit commander 1955>

Col Landon McConnell base/unit commander 1956>
Lt Col Cy Perkins   unit commander 1957>
Lt Col Walter Rosenfield   unit commander 1958>

Detachment A
(cover designation Weather Reconnaissance Squadron (Provisional) - 1 (WRSP-1) established at Watertown Strip January 1956, to Lakenheath April 1956, to Wiesbaden June 1956, to Giebelstadt October 1956, closed November 1957
Col Fred McCoy   unit commander 1956>
Col Marion Mixson   unit commander July 1957>
Bob King   executive officer 1956>
Tom Clendenan   executive officer 1957>
Maj Phil Karras   operations officer 1956>
Lt Col Dale Garrison   operations officer 1957>

Detachment B
(cover designation Weather Reconnaissance Squadron (Provisional) - 2 (WRSP-2) established at Watertown Strip spring 1956, to Incirlik (Adana) August 1956 (where unit also known as Det 10-10, TUSLOG), closed June 1960
Col Ed Perry   unit commander 1956>
Col Stan Beerli   unit commander November 1957>
Col William Shelton   unit commander August 1959>
John Parangosky   executive officer 1956>
Dick Newton   executive officer 1958>
Jim Sether   executive officer 1959>
Lt Col Cy Perkins   operations officer 1956>
Lt Col Chet Bohart   operations officer 1956>
Maj Ray Sterling   operations officer November 1957>
Maj Don Scherer   operations officer 1959>

Detachment C
(cover designation Weather Reconnaissance Squadron (Provisional) - 3 (WRSP-3) established at Watertown Strip August 1956, to Atsugi March 1957, closed June 1960
Col Stan Beerli   unit commander 1956>
Col Marion Mixson   unit commander November 1957>
Bill Crawford   executive officer 1957>
Werner Weiss   executive officer November 1957>
Lt Col Robert Larkin   operations officer 1956>
Lt Col Chuck Quinette   operations officer November 1957>

Pilots

Barry Baker	Det C throughout
Jim Barnes	Det C, to Det B 1957
Tom Birkhead	Det B, to USAF 1957, to test unit 1958
Mike Bradley	RAF, to Det B 1958
Howard Carey	trained with Det B but assigned to Det A, killed 17 September 1956
Jim Cherbonneaux	trained with Det C but assigned to Det B, to test unit 1957
Tom Crull	Det C throughout
David Dowling	RAF, to Det B 1958
Glen Dunaway	Det A, to USAF 1957, to test unit 1958, to Det B 1960
Buster Edens	Det B, to Det C 1957
Bob Ericson	Det C, to Det B 1957
Bill Hall	Det B, to USAF 1957
E.K. Jones	Det B, to Det C 1957
Marty Knutson	Det A, to Det B 1957
Jake Kratt	Det A, to USAF 1957, to test unit 1958, to Det B 1960
John MacArthur	RAF, to Det B 1958
Bill McMurry	Det B, to Det C 1957
Carl Overstreet	Det A, to USAF 1957
Frank Powers	Det B throughout
Al Rand	Det C throughout
Robbie Robinson	RAF, to Det B 1958
Lyle Rudd	Det C throughout
John Shinn	Det C, to Det B 1957
Al Smiley	trained with Det C but assigned to Det A, to USAF 1957
Sammy Snyder	Det B, to Det C 1957
Hervey Stockman	Det A, to USAF 1957
Carmine Vito	Det A, to test unit 1957, to Project HQ as ops officer 1959

The above is a complete list of those who trained and deployed as operational U-2 pilots to the CIA detachments. Some unit commanders and operations officers also flew the U-2 on training and ferry flights. The following did not complete training: Jim Allison (Det A); Frank Grace (Det C, killed 31 Aug 1956); Wilbur Rose (Det B, killed 15 May 1956); Bill Strickland (Det B); Chris Walker (RAF, killed 8 July 1958). The following completed training, but did not deploy: Bunny Austin (RAF, 1959), Byron Cox (RAF, 1959).

Appendix 3:
USAF Organization and Key Personnel

1007th Air Intelligence Support Group, HEDCOM
was the small office in the Pentagon which provided USAF liaison to Project HQ
Col Ozzie Ritland   December 1954>
Col Russ Berg   September 1955>
Col Leo Geary   May 1956>

4070th Support Wing
This was the 'cover' designation used by the group which was established by SAC
in later 1955 to provide U-2 training and advice to the CIA at Watertown Strip. It
was officially based at March AFB, CA.
Commander
Col Bill Yancey
Pilots
Louis Garvin, Hank Meierdierck, Bob Mullin, Phil Robertson, Louis Setter

4080th Strategic Reconnaissance Wing (SRW)
Activated at Turner AFB, GA on 1 April 1956, reporting to 2nd Air Force, Strategic
Air Command. The first mission was to operate the RB-57Ds of Project Black
Knight, for which purpose the 4025th SRS was assigned to the wing in May 1956.
The second mission was to operate the U-2s ordered by SAC under Project Dragon
Lady, and the 4028th SRS was created for this purpose. Prior to receiving its U-2s,
the 4080th wing moved to Laughlin AFB, TX during February-April 1957. A second U-2 squadron, the 4029th SRS, was assigned but never equipped.
Commanders
Col Gerald Johnson   May 1956>
Col Hub Zemke   April 1957>
Bg Gen Austin Russell   November 1957>

Col Andrew Bratton   December 1958>
Deputies for Operations
Lt Col Howard Cody   May 1956>
Lt Col John Moore   September 1957>
Col Howard Shidal   November 1957>
Lt Col John Boynton   October 1958>
Col John Swofford   January 1959>
4028th SRS Commanders
Lt Col Jack Nole   July 1956>
Lt Col Hayden Curry   April 1958>
4028th SRS U-2 Pilots
listed in alphabetical order by year of checkout, until May 1960. The list includes
wing staff pilots and others who did not achieve operational status.
1956: Jack Nole, Joe Jackson, Floyd Herbert, Hank Nevett, Howard Cody
1957: Nat Adams, Ken Alderman, Skip Alison, Rudy Anderson, Dick Atkins, Snake
Bedford, Tony Bevacqua, Jim Black, Warren Boyd, John Campbell, Alfred Chapin,
Roger Cooper, Harry Cords, Buzz Curry, Ed Dixon, Marv Doering, Ed Emerling,
Bobbie Gardiner, Jack Graves, Ray Haupt, Pat Halloran, Roger Herman, Joe King,
Steve Heyser, Cozy Kline, Bennie Lacombe, Dick Leavitt, Buck Lee, Ford Lowcock,
Earl Lewis, Wes McFadden, Richard McGraw, John McElveen, Ed Perdue, Bob
Pine, Jim Qualls, Jim Sala, Leo Smith, Scott Smith, Roy St Martin, Frank Stuart,
Mike Styer, Forrest Wilson.
1958: Adrian Acebado, Buddy Brown, John Boynton, Dick Callahan, Bob Ginther,
Paul Haughland, Don James, Bo Reeves, Bill Rodenbach, Austin Russell, Ken Van
Zandt, Bob Wood.
1959: Andrew Bratton, Jack Carr, Deke Hall, Ron Hedrick, Floyd Kifer, TJ Jackson
Jr, Kenneth McAslin, Gerry McIlmoyle, Harold Melbratten, Raleigh Myers,
Ellsworth Powell, Bob Powell, Dick Rauch, J.B. Reed, Bob Schueler, Bill Stickman,
Chuck Stratton.
1960: Don Crowe, Rex Knaak, Tony Martinez, Bob Spencer, Bob Wilke.

Special Projects Branch, USAF Flight Test Center (AFFTC)
this unit was activated at Edwards AFB in early 1958 for flight tests of the U-2 with
infrared sensors for aircraft and (later) missile detection.
Commanders
Major Robert Carpenter   1958-59
Major Harry Andonian   1960
Pilots
Donald Evans, Mervin Evenson, Pat Hunerwadel, Robert Jacobsen, Bud Knapp,
Rial Lowell, Lachlan Macleay, Philip Smith
Observers
James Eastlund, Bill Frazier, Ray Oglukian

Appendix 4:
Lockheed Key Personnel

Original Design Group, January1955>
Kelly Johnson   chief engineer
Dick Boehme   project engineer
Carl Allmon (loft), Ed Baldwin (fuselage), Bill Bissell (wing), Vern Bremberg (hydraulics), Lorne Cass (basic loads), Bob Charlton (drawings, manuals), Henry Combs (stress), Doug Cone (air conditioning), Royal Dow (wing), Chan Engelbry (aerodynamics), Leroy English (fuselage), George Ellison (control system), Elmer Gath (propulsion), Cornelius Gardner (landing gear), Leon Gavette (ground handling equipment), Pete Gurin (static test), John Henning (static test), Richard Kruda (stress), Alvin Jensen (loft), Bob Kelly (aileron), Ray Kirkham (fuselage), Rod Kriemendahl (horizontal tail), Lutz (ststic test), Ed Martin (sensors), Ray McHenry (stress), Sam Murphy (electrical), Herb Nystrom (vertical tail), Jack Painter (fuselage), Dave Robertson (fuel system), Al Robinson (weights), Cliff Rockel (electrical), Vic Sorenson (control system), Bob Wiele (wing), Dan Zuck (cockpit).

Test Pilots (listed in order of their first U-2 flights)
1955: Tony LeVier
1955>: Bob Matye, Ray Goudey, Bob Sieker (killed 4 April 1957), Bob Schumacher
1959: Ted Limmer

Flight Test Engineers
Ernie Joiner (chief), Paul Deal, Glen Fulkerson, Bob Klinger, P.E. Smith, Mel Yoshi (1957>)

Project and Field Service Managers
Art Bradley   project manager October 1959>
Frank Bertelli   field service at Det C >1957, then test unit

Buzz Johnson	field service at Det B >1957, and Det B 1959>
Dorsey Kammerer	shop manager, Watertown
Ed Martin	project manager January 1958-October 1959
Bob Murphy	field service at Det C >1957, again in 1958, then at Det B >1959
Howard Slack	field service at Det A
Art Westlund	production manager, Oildale

Appendix 5:
Aircraft Production and Service Histories

NOTE: The primary identifier for each airframe was the "Article number", equivalent to a manufacturers' construction number. The CIA and Lockheed continued to use these numbers, even after USAF serial numbers were allocated in early 1956. To enhance the cover story, CIA aircraft routinely carried a different, three-digit NACA number. These appear to have been allocated at random, with the same number applied to different aircraft from time to time: Det A used various NACA numbers from 160 through 199; Det B appears to have used numbers from 300 through 310, and Det C some numbers in the 400 series. No comprehensive listing of these NACA numbers appears to have survived.

Although the USAF only ordered 29 aircraft initially, Fiscal Year 1956 military serial numbers were also allocated retrospectively to the aircraft previously ordered by the CIA, with the exception of the prototype Article. In 1957, the USAF ordered a 30th aircraft. Confusingly, this was given the same military serial number as had previously been allocated to a CIA aircraft (Article 357) which crashed in December 1956. When the USAF ordered another five aircraft in 1958, they were also given Fiscal 1956 serial numbers, instead of Fiscal 1958 ones.

Transfers of aircraft between the CIA and USAF and vice versa began in 1957, and were a constant feature thereafter. As an aid to quick identification, maintenance personnel could add 333 to the Article number, to obtain the last three digits of the USAF serial number (with the exception of Articles 390-395). Incidentally, the Skunk Works referred to the CIA as "Customer One" and the USAF as "Customer Two".

All previously-published lists of U-2 production have contained errors. Although this listing is compiled largely from unofficial sources, the author believes that it is accurate. Data up to June 1960 is presented here, reflecting the timespan covered in this book.

Article 341 delivered to Groom Lake on 24 July 1955 and first flown on 1 August. Remained at the test site for development work until it crashed in open desert some 50 miles north of the test site during a Project Rainbow (stealth) test flight on 4 April 1957. Lockheed test pilot Robert Sieker was killed.

Article 342/56-6675 delivered to Groom Lake on 11 September 1955 and allocated to flight tests of the camera systems. Damaged in landing accident 21 March 1956 by CIA pilot Carmine Vito, but repaired and deployed to Det B at Adana, Turkey in August 1956. Returned to US in early 1957 and went to Det C at Atsugi, Japan during May 1957. By early 1959, was at North Base for conversion to U-2C. First flew as the prototype U-2C on 13 May 1959. Remained there as the U-2C test aircraft, and was used for the flight tests in June 1960 which attempted to determine what had gone wrong on the failed overflight mission of 1 May 1960.

Article 343/56-6676 delivered to Groom Lake on 16 October 1955. Used for test and training purposes throughout 1956, but damaged in early 1957. After repairs, became the testbed for long lens "C" camera June-November 1957 and is then believed to have deployed to Det C in Japan. Returned to North Base August – October 1958. "C" camera provisions removed in 1959.

Article 344/56-6677 delivered to Groom Lake on 20 November 1955. Engaged in test and development work through most of 1956. Repaired after belly landing on lakebed 1 June 1956 when CIA pilot Bill Strickland ran out of fuel on final approach. Probably deployed overseas during 1957, possibly to Det C under Project Rainbow (stealthed). At North Base for IRAN (Inspect and Repair As Necessary) August-October 1958, and again February-May 1959.

Article 345/56-6678 delivered to Groom Lake on 16 December 1955. Used in the training program for the first two groups of CIA pilots. Crashed at the test site on 15 May 1956 when CIA pilot Wilbur Rose stalled while trying to release a jammed pogo.

Article 346/56-6679 delivered to Groom Lake on 13 January 1956. Used for training flights until sent on the initial Det A deployment to UK in late April 1956. Crashed on 17 September 1956 during climb-out from Wiesbaden, Germany, killing CIA pilot Howard Carey.

Article 347/56-6680 delivered to Groom Lake on 8 February 1956. Used for training flights until sent on the initial Det A deployment in late April 1956. Marked with tail number NACA 187. Flew the first operational mission over Eastern Europe on 20 June 1956, and the first two missions over the USSR on 4 and 5 July. Returned to US in October 1957 when Det A was closed down and transferred to USAF 4080th SRW at Laughlin AFB. At factory for IRAN in spring 1960.

Article 348/56-6681 delivered to Groom Lake on 5 March 1956. With Det A on initial deployment in late April 1956. Returned to US by November 1957 and retained at North Base through January 1959 for test and development work. Served as the prototype for the System 6 ELINT installation. Transferred to USAF 4080th SRW in mid-1959 and modified to carry the APQ-56 radar.

Article 349/56-6682 delivered to Groom Lake on 29 March 1956. With Det A on initial deployment in late April 1956, and was one of the three CIA aircraft equipped with APQ-56 side-looking radar. Probably one of first four aircraft to be plumbed for slipper tanks during 1957. Project Buckhorn (stealth) tests, June 1958. Deployed to Det B during 1958. While serving with Det C on 5 April 1960, ran out of fuel on return from a China overflight and was damaged when Bill McMurray landed in a Thai rice paddy.

Article 350/56-6683 delivered to Groom Lake on 24 April 1956. Deployed to Det B in August 1956 equipped with APQ-56 radar. Returned to US in fall 1957 and transferred to USAF 4080th SRW, retaining the side-looking radar.

Article 351/56-6684 delivered to Groom Lake on 18 May 1956. Third CIA aircraft equipped with APQ-56 side-looking radar. Possibly retained at the test site throughout 1956. Converted to Project Rainbow radar-stealthy configuration in early 1957 and deployed in July to Det B, which employed it on USSR overflights from Lahore during August. Transferred to Det A at Giebelstadt in October 1957 from where it flew one overflight and one peripheral ELINT flight to the Kola Peninsula region of USSR. Became the third conversion to U-2C by July 1959 and deployed to Det B the following month.

Article 352/56-6685 delivered to Groom Lake on 13 June 1956. Deployed to Det B in August 1956. Believed to have returned to factory for slipper tank installation in 1958. Converted to U-2C by September 1959 and probably deployed to Det C by year end, 1959.

Article 353/56-6686 was delivered to Groom Lake on 6 July 1956. Deployed to Det B in August 1956. Remained a U-2A model, but little else is known about this aircraft's Agency career pre-1961.

Article 354/56-6687 was delivered to Groom Lake on 27 July 1956. Crashed there on 31 August 1956 when CIA pilot Frank Grace became disoriented during a night takeoff and stalled.

Article 355/56-6688 was delivered to Groom Lake on 16 August 1956. Damaged on landing 29 August 1956, but not repaired until December. Believed to have been operated by Det B during 1958 – 59. Was probably prepared for System 7 installation while at North Base in November 1959. Redeployed to Det B thereafter, still as U-2A.

Article 356/56-6689 was delivered to Groom Lake on 5 September 1956. Deployed to Det B by April 1957. Transferred to USAF 4080th SRW in November 1957.

Article 357/56-6690 was delivered to Groom Lake on 21 September 1956. This aircraft was probably awaiting deployment with Det C when it crashed onto an Indian reservation in Arizona on 19 December 1956. CIA pilot Bob Ericson became hypoxic due to a fault in the oxygen feed and lost control, but managed to recover in time to bailout at medium altitude. See also Article 390.

Article 358/56-6691 was delivered to Groom Lake on 8 October 1956. Deployed to Det C in March 1957. At North Base for test/development work from September 1958, and flew the stability and control tests requested by USAF in spring 1959. Was the second conversion to U-2C by July 1959 and deployed to Det B in August.

Article 359/56-6692 was delivered to Groom Lake on 22 October 1956. Deployed to Det A in November 1956. Returned to US when Det A closed in November 1957. Deployed to Det C in June 1958. Returned to North Base in September 1959.

Article 360/56-6693 was delivered to Groom Lake on 5 November 1956. Deployed to Det C in March 1957. Flew Det C's first USSR overflight on 8 June 1957. Returned to North Base by September 1957 where it was engaged in test and development work until May 1959. Converted to U-2C in August 1959 and deployed again to Det C in September. Marked with tail number 449. Damaged in belly landing at Fujisawa airfield on 24 September 1959 after CIA pilot Tom Crull ran out of fuel on approach to Atsugi. After repair at North Base, deployed to Det B in March 1960. Shot down near Sverdlovsk, USSR on 1 May 1960, when CIA pilot Francis G. Powers became a guest of the KGB.

Article 361/56-6694 was the first aircraft built at Oildale under the USAF contract. Delivered to Groom Lake in September 1956. Used there for training the initial 4080th SRW pilot cadre until flown to Laughlin AFB in June 1957. Undamaged during deadstick landing into a Texas ranch by Capt Jim Qualls on 21 September 1957, but crashed during subsequent functional check flight on 26 September. After losing control during an uncommanded flap deployment, Col Jack Nole performed the highest successful bail out from an aircraft, within 25 miles of Laughlin AFB.

Article 362/56-6695 was delivered to Groom Lake in November 1956. Used there for training the initial 4080th SRW pilot cadre until flown to Laughlin AFB in June 1957. Remained with 4080th SRW thereafter.

Article 363/56-6696 was delivered to Groom Lake in December 1956. Damaged during USAF pilot training in an early 1957 landing accident by Lt Tony Bevacqua. Repaired by July and assigned to the 4080th SRW at Laughlin AFB.

Article 364/56-6697 was delivered to Groom Lake in January 1957. Used there for training the initial 4080th SRW pilot cadre until flown to Laughlin AFB in June 1957. Stalled and crashed on final approach to Laughlin on 6 August 1958, killing Lt Paul Haughland.

Article 365/56-6698 was delivered to Groom Lake in January 1957. The first USAF U-2 to carry the genuine weather reconnaissance package. Flown to Laughlin AFB in June 1957 for continued service with the 4080th SRW. Deployed to Eielson AB for the first Toy Soldier sampling flights January-March 1958. Crashed 23 miles southwest of Tucumcari, New Mexico on 9 July 1958, after suspected oxygen and/ or autopilot problems, killing Capt Alfred Chapin.

Article 366/56-6699 was delivered to Groom Lake in February 1957. Flown to Laughlin AFB in June for continued service with the 4080th SRW. Crashed only days later on the 28th June during a high altitude training flight over New Mexico, killing Lt. Leo Smith. Cause was undetermined.

Article 367/56-6700 was delivered to Groom Lake in February 1957 under the USAF contract but appears to have been transferred to the CIA by June when it was ferried from there to North Base. Probably received slipper tanks by the end of 1957 and deployed overseas but little is known of this aircraft until it was transferred to the 4080th SRW in fall 1960.

Article 368/56-6701 was delivered to Groom Lake in March 1957. Retained at Edwards after transfer there in June 1957. Assigned to Special Projects Branch, AFFTC there by the end of 1959, and used in a number of research projects thereafter.

Article 369/56-6702 was delivered to Groom Lake in March 1957. Possibly the first U-2 to land at Laughlin AFB on 10 June 1957 when the 4080th SRW transferred Dragon Lady operations to there. Crashed only 17 days later, on 28 June, when Lt Ford Lowcock lost control at low level over Del Rio and became the first USAF U-2 fatality.

Article 370/56-6703 was delivered to Groom Lake in April 1957. Transferred to Laughlin AFB for further service with 4080th SRW in June 1957. Deployed to Eielson AB for the first Toy Soldier sampling flights January-March 1958, and again September-November 1958. Modified to USAF's dedicated SIGINT configuration (Systems 1/3/4) in 1959.

Article 371/56-6704 was delivered to Groom Lake in April 1957. Transferred to Laughlin AFB for further service with 4080th SRW in June 1957. Crashed on base leg to Laughlin on the night of 28 November 1957 when Capt Benedict Lacombe lost control. He was killed after an unsuccessful bailout.

Article 372/56-6705 was delivered to Groom Lake in April 1957. The prototype "hard-nose" sampling aircraft (U-2A-1 model) for the HASP project. Therefore frequently deployed by 4080th SRW to Ramey, Plattsburgh and Minot AFBs from October 1957 onwards.

Article 373/56-6706 was delivered to Groom Lake in May 1957. Transferred to Laughlin AFB for further service with 4080th SRW in June 1957. Deployed to Eielson for Toy Soldier sampling flights September-November 1958.

Article 374/56-6707 was delivered to Groom Lake in May 1957. Transferred to Laughlin AFB for further service with 4080th SRW in June 1957. Deployed to Eielson AB for the first Toy Soldier sampling flights January-March 1958, and again September-November 1958, September-October 1959. Modified to USAF's dedicated SIGINT configuration (Systems 1/3/4) in 1959. Deployed again to Eielson March-May 1960.

Article 375/56-6708 was delivered to Groom Lake or North Base in June 1957 and thence to 4080th SRW at Laughlin AFB. Equipped with the APQ-56 radar.

Article 376/56-6709 was delivered to Groom Lake or North Base in May 1957. The first USAF aircraft to be configured for dedicated SIGINT (Systems 1/3/4 plus ASN-6 RADAN), and therefore retained at North Base through November 1957 for test flights. Assigned to 4080th SRW at Laughlin AFB thereafter.

Article 377/56-6710 was delivered to North Base in June 1957. Modified to carry an observer in Q-bay and flying as such by early January 1958. Assigned to the Special Projects Branch, AFFTC and IR search equipment possibly added. Destroyed in an accident at Edwards on 11 September 1958 which killed Capt Pat Hunerwadel.

Article 378/56-6711 was delivered to North Base in July 1957. Although built under the USAF contract, this aircraft was apparently assigned to the CIA. One of the first aircraft equipped with slipper tanks. Frequently at North Base throighout 1958-60. Marked as 'NASA 55741', this aircraft was put on static display at Edwards main base for the media's benefit on 6 May 1960, after the shooting-down of Article 360 six days earlier.

Article 379/56-6712 was delivered to North Base in July 1957 and subsequently assigned to the 4080th SRW at Laughlin AFB.

Article 380/56-6713 was delivered to North Base in July 1957 and subsequently assigned to the 4080th SRW. Configured for dedicated SIGINT (Systems 1/3/4 plus ASN-6 RADAN). Crashed near Wayside, Texas on 8 July 1958, killing RAF Sqn Ldr Chris Walker. Oxygen and/or autopilot problems were suspected causes.

Article 381/56-6714 was delivered to North Base in August 1957 in "hard-nose" (U-2A-1) sampling configuration for Project HASP. Subsequently deployed by the 4080th SRW to Ramey AFB, and Ezeiza, Argentina in 1958-59.

Article 382/56-6715 was delivered to North Base in August 1957 in "hard-nose" (U-2A-1) sampling configuration for Project HASP. Subsequently deployed by the 4080th SRW to Ramey and Minot AFBs, and Ezeiza, Argentina in 1958-60.

Article 383/56-6716 was delivered to North Base in September 1957 in "hard-nose" (U-2A-1) sampling configuration for Project HASP. Subsequently deployed by the 4080th SRW to Plattsburgh AFB and Ezeiza, Argentina in 1958-60. Undamaged during deadstick landing on frozen lake at Prince Albert, Saskatchewan, Canada on 15 March 1960 by Capt Roger Cooper.

Article 384/56-6717 was delivered to North Base in September 1957 in "hard-nose" (U-2A-1) sampling configuration for Project HASP. Subsequently deployed by the 4080th SRW to Plattsburgh and Minot AFBs and Ezeiza, Argentina in 1958-60.

Article 385/56-6718 was delivered to North Base in September 1957 in "hard-nose" (U-2A-1) sampling configuration for Project HASP. Subsequently deployed by the 4080th SRW to Plattsburgh AFB and Ezeiza, Argentina in 1958-60.

Article 386/56-6719 was delivered to North Base in October 1957 and subsequently assigned to the 4080th SRW. Configured for dedicated SIGINT (Systems 1/3/4 plus ASN-6 RADAN).

Article 387/56-6720 was delivered to North Base in October 1957 and subsequently assigned to the 4080th SRW. Configured for dedicated SIGINT (Systems 1/3/4 plus ASN-6 RADAN).

Article 388/56-6721 was delivered to North Base in October 1957 and subsequently assigned to the 4080th SRW. Damaged in belly landing at Cortez, Colorado during training flight on 5 August 1959. Modified during repair to house observer and IR search equipment in Q-bay, flown in December 1959 and assigned to the Special Projects Branch/AFFTC at Edwards. Unofficially referred to as a U-2B model, but later formally designated a U-2D.

Article 389/56-6722 was delivered to North Base in November 1957. Was the first U-2 to be modified to accept the IR search equipment in the Q-bay, but remained a single-seat aircraft with no provision for an observer. Delivered to the Special Projects Branch/AFFTC at Edwards in March 1958.

Article 390/56-6690 was the 30th and last aircraft produced under the first USAF contract, and was the second aircraft to be assigned the serial number 56-6690. Delivered to North Base in November or December 1957. Subsequently assigned to the 4080th SW.

Article 391/56-6951 was the first of five aircraft produced under the second USAF contract. First flown at North Base in December 1958 or January 1959. Assigned to the 4080th SW, and deployed to Eielson AB for Toy Soldier sampling and Congo Maiden peripheral photo flights March-May 1959, September-October 1959, and March-May 1960.

Article 392/56-6952 was flying at North Base in January 1959. Assigned to the 4080th SW, and deployed to Eielson AB for Toy Soldier sampling and Congo Maiden peripheral photo flights March-May 1959.

Article 393/56-6953 was flying at North Base in February 1959. Assigned to the 4080th SW, and deployed to Eielson AB for Toy Soldier sampling and Congo Maiden peripheral photo flights March-May 1959, September-October 1959, and March-May 1960.

Article 394/56-6954 was flying at North Base in March 1959. Was built with the modified Q-bay housing an observer and IR search equipment. Assigned to the Special Projects Branch/AFFTC at Edwards. Unofficially referred to as a U-2B model, but later formally designated a U-2D.

Article 395/56-6955 was flying at North Base in February 1959. Assigned to the 4080th SRW and carried the APQ-56 radar.

Index